ACS SYMPOSIUM SERIES **684**

Enzymes in Polymer Synthesis

Richard A. Gross, EDITOR
University of Massachusetts

David L. Kaplan, EDITOR
Tufts University

Graham Swift, EDITOR
Rohm & Haas Company

Developed from a symposium sponsored by
the Division of Polymeric Materials: Science and Engineering
at the 211th National Meeting of the American Chemical Society,
New Orleans, LA,
March 24–28, 1996

American Chemical Society, Washington, DC

Library of Congress Cataloging-in-Publication Data

Enzymes in polymer synthesis / Richard A. Gross, David L. Kaplan, Graham Swift.

p.   cm.—(ACS symposium series, ISSN 0097–6156; 684)

"Developed from a symposium sponsored by the Division of Polymeric Materials: Science and Engineering at the 211th National Meeting of the American Chemical Society, New Orleans, LA, March 24–28, 1996."

Includes bibliographical references and indexes.

ISBN 0–8412–3543–0

1. Enzymes—Biotechnology—Congresses. 2. Polymerization—Congresses. 3. Polymers—Congresses.

I.Gross, Richard A., 1957– . II. Kaplan, David, 1953– . III. Swift, Graham, 1939– . IV. American Chemical Society. Division of Polymeric Materials: Science and Engineering. V. American Chemical Society. Meeting (211th: 1996: New Orleans, La.) VI. Series.

TP248.65.E59E625   1998
668.9—dc21   97–38746
                                                            CIP

This book is printed on acid-free, recycled paper.

PRINTED IN THE UNITED STATES OF AMERICA

Advisory Board

Foreword

THE ACS SYMPOSIUM SERIES was first published in 1974 to provide a mechanism for publishing symposia quickly in book form. The purpose of the series is to publish timely, comprehensive books developed from ACS sponsored symposia based on current scientific research. Occasionally, books are developed from symposia sponsored by other organizations when the topic is of keen interest to the chemistry audience.

Before agreeing to publish a book, the proposed table of contents is reviewed for appropriate and comprehensive coverage and for interest to the audience. Some papers may be excluded in order to better focus the book; others may be added to provide comprehensiveness. When appropriate, overview or introductory chapters are added. Drafts of chapters are peer-reviewed prior to final acceptance or rejection, and manuscripts are prepared in camera-ready format.

As a rule, only original research papers and original review papers are included in the volumes. Verbatim reproductions of previously published papers are not accepted.

ACS BOOKS DEPARTMENT

Contents

Preface vii

INTRODUCTION

1. **Enzymes in Polymer Science: An Introduction** 2
   David L. Kaplan, Jonathan Dordick, Richard A. Gross,
   and Graham Swift

SYNTHESIS

2. **Enzymes for Polyester Synthesis** 18
   Apurva K. Chaudhary, Eric J. Beckman, and Alan J. Russell

3. **Enzymatic Polymerization for Synthesis of Polyesters
   and Polyaromatics** 58
   Shiro Kobayashi and Hiroshi Uyama

4. **Enzyme-Catalyzed Ring-Opening Polymerization of Four-Membered
   Lactones: Preparation of Poly(β-propiolactone) and Poly(β-malic acid)** 74
   Shuichi Matsumura, Hideki Beppu, and Kazunobu Toshima

5. **Monomer and Polymer Synthesis by Lipase-Catalyzed Ring-Opening
   Reactions** 90
   Kirpal S. Bisht, Lori A. Henderson, Yuri Y. Svirkin, Richard A. Gross,
   David L. Kaplan, and Graham Swift

6. **Solvent–Enzyme–Polymer Interactions in the Molecular-Weight Control
   of Poly(m-cresol) Synthesized in Nonaqueous Media** 112
   Madhu Ayyagari, Joseph A. Akkara, and David L. Kaplan

7. **A Biocatalytic Approach to Novel Phenolic Polymers and Their
   Composites** 125
   Sukanta Banerjee, Premchandran Ramannair, Katherine Wu,
   Vijay T. John, Gary McPherson, Joseph A. Akkara,
   and David L. Kaplan

v

8. Hydrolysis of the Dimethyl Ester of *N*-Succinylphenylalanine, a Model of Polyesteramides, in the Presence of Papain: Kinetic Study and Computer Simulation 144

    C. David, F. G. O. Lefebvre, R. Brasseur, and M. Vanhaelen

POLYMER MODIFICATIONS

9. Regioselective Enzymatic Transesterification of Polysaccharides in Organic Solvents 167

    Ferdinando F. Bruno, Jonathan S. Dordick, David L. Kaplan, and Joseph A. Akkara

10. Biocatalytic Modification of Alginate 175

    Donal F. Day, R. D. Ashby, and J. W. Lee

11. Enzymatic Modification of Chitosan by Tyrosinase 188

    Joseph L. Lenhart, Mahesh V. Chaubal, Gregory F. Payne, and Timothy A. Barbari

12. Chemoenzymatic Synthesis and Modification of Monomers and Polymers 199

    Helmut Ritter

Author Index 209

Affiliation Index 209

Subject Index 210

Preface

THE IDEA THAT NATURE HAS MUCH TO TEACH the chemical industry about the control and efficiency of chemical transformations is increasing in acceptance. As the industry moves toward sustainable development, some of the major goals are to limit disposal requirements by limiting the production of unwanted side products, to lower energy requirements and costs, to control polymerization for both structure and composition, to limit emissions, to use renewable resources wherever possible, and to design biodegradable polymers where environmental disposal is required. Many of these goals may be met by a better understanding and wider application of enzymatic processes for both the synthesis and degradation of polymers. Already, there are important examples of enzyme applications on a high-volume scale in several industries, such as the use of glucose isomerase in the food inustry; cellulase in textile finishing; and lipases, cellulases, and proteases in detergents. There are also extensive data on the enzymatic synthesis of chiral intermediates in the pharmaceutical industry.

This book was developed from a symposium at the 211th National Meeting of the American Chemical Society, sponsored by the ACS Division of Polymeric Materials: Science and Engineering, Inc., titled "Biocatalysis in Polymer Chemistry", in New Orleans, Louisiana, March 24–28, 1996. This book is believed to be the first to address the significance of enzymes in polymer science. Many leading scientists in the field have contributed chapters on different aspects of enzyme-based polymer science, including a broad overview and an introduction to the subject, and chapters on enzyme synthesis from monomers, specific modification of polymers, and enzymatic degradation with the intent of gaining a fundamental understanding of mechanisms and developing models.

The editors of this book are convinced that the use of enzymes in polymer science is on the verge of rapid expansion. This chang is due to recent advances in enzyme isolation and enzyme engineering, better analytical tools for monitoring reactions, and improved modeling techniques. Opportunities for the development of new technologies are legion, and the rewards for everyone are great, both technically and environmentally. These are the very goals of a sustainable chemical industry.

RICHARD A. GROSS
Department of Chemistry
University of Massachusetts—Lowell
One University Avenue
Lowell, MA 01854

DAVID L. KAPLAN
Biotechnology Center
Department of Chemical Engineering
4 Colby Street
Tufts University
Medford, MA 02155

GRAHAM SWIFT
Research Laboratories
Rohm and Haas Company
727 Norristown Road
Spring House, PA 19477–0904

September 12, 1997

INTRODUCTION

Chapter 1

Enzymes in Polymer Science: An Introduction

David L. Kaplan[1], Jonathan Dordick[2], Richard A. Gross[3], and Graham Swift[4]

[1]Biotechnology Center, Department of Chemical Engineering, Tufts University,
4 Colby Street, Medford, MA 02155
[2]Department of Chemical and Biochemical Engineering and Center
for Biocatalysis and Bioprocessing, University of Iowa, Iowa City, IA 52242
[3]Department of Chemistry, University of Massachusetts—Lowell,
Lowell, MA 01854
[4]Rohm and Haas Company, 727 Norristown Road, Spring House, PA 19477

Enzymes are important catalysts in a wide range of reactions because of their catalytic rates, specificity and function under mild conditions. These features have led to extensive studies of enzyme structure and function, particularly with small molecules. These studies have also led to the use of enzymes for many industrially-relevant applications. In recent years the benefits of enzymes in various reactions in polymer science also have been investigated. These studies have been catalyzed by at least two major interests: (a) issues related to 'green chemistry' and (b) the discovery that enzymes function in novel environments such as environmental extremes and organic solvents. These interests have spurred a new generation of activity to explore the role of these protein-based catalysts in a wide range of reactions related to polymer synthesis and polymer modification. This volume captures some of these activities and documents new directions and advances in the use of enzymes in polymer science.

1. Introduction

1.1. Traditional Areas of Application Using Enzymes-

Enzymes have found traditional use in biochemical studies, molecular biology and related scientific explorations. In addition, high volume industrial uses for enzymes have focused on food processing (e.g., glucose isomerase), detergent additives (e.g., lipases and proteases) and textile finishing (e.g., cellulase). Additional opportunities are being realized with the exploration of new food processing enzymes, such as pectinase; new detergent enzymes for liquids and powders; continued interest in modifying or improving cellulases for textiles; and applications for ligninases, cellulases, hemicellulases and xylanases to help pulping operations as replacements for chlorine-based chemistries. Many of these industrial uses involve hydrolytic reactions in aqueous systems as the primary mode of action. There also are extensive data on the use of enzymes for the synthesis or modification of small substrate molecules, from chiral pharmaceutical intermediates to modified sugars (e.g., see Margolin, 1993; Wang *et al.*, 1995). Future opportunities exist for many new applications for enzyme-derived materials, including coatings and films, separation membranes, emulsifiers and dispersants, biomaterials, composites, photonics and electronics.

1.2. Enzymes and Polymer Synthesis - New Directions

This volume and this introductory chapter will focus primarily on the use of biocatalysis in the field of polymer science. We will address the methods being used to synthesize or modify polymers through biocatalytic reactions. The level of activity in this field has been growing in recent years, however, in comparison to studies with small molecules, there remains only limited research in the polymer field. It is hoped that this volume will help identify some of the exciting opportunities in the emerging area of enzymes in polymer science and stimulate additional thought and exploration. Unlike the traditional uses for enzymes that often involve hydrolysis, many of the new directions in polymer science focus on condensation, ring opening and controlled free-radical polymerizations.

1.3. Biomimetics (Mimicking Biology)

It is perhaps worth noting at the onset of this volume that the biological world builds and modifies all of its polymers using enzymes. Thus, there are many lessons to be learned from the integration of the biological sciences with polymer science to aide in polymer synthesis, chemistry and processing, leading to new and improved materials. In biology, enzymes function in the construction and degradation of polysaccharides, proteins, polyphenols and polynucleic acids. These reactions occur in aqueous media, in lipophilic membranes or at interfaces between these two phases. It is ironic that we are only beginning to employ these features to help modulate enzymatic reactions *in vitro* to generate new polymers. In the biological world, the synthesis of biopolymers is essential to survival, despite the metabolic cost to the organism to carry out the process. Importantly, the ability of an organism to use enzymes to degrade and reuse biopolymers for metabolic needs is also a critical function in closing the materials life cycle loop. This is another biological process that we are only beginning to emulate with synthetic biodegradable polymers. Thus, there remains a great deal to be learned using the paradigm of enzymatic synthesis and degradation of polymers in the biological context. This approach is also being expanded with new insights into nonprotein-based catalytic systems, such as ribozymes.

2. Key Factors to Consider in Enzyme-Based Polymerization Studies

- Control of Polymer Structure
- Ease and Flexibility of Reactions
- Green Chemistry

The use of biocatalysts in polymer synthesis and modification is driven by many factors. These can be summarized as follows:

2.1. Control of Polymer Structure

Enzymes exhibit enantioselectivity, preferentially reacting with only one stereoisomer, and regioselectivity, preferentially reacting with only one site on a molecule despite multiple sites of potential reaction. These features are ingrained in the world of biology, where for the most part control of stereochemistry and regioselectivity in enzyme-based reactions is essential for function. For example, only L-amino acids are used in enzymatic polymerization reactions to build proteins and usually only the α-carbons of the amino acids are coupled by peptide synthases to form covalent peptide bonds to hold the amino acids together. In a similar fashion, mostly D-sugars are utilized to build polysaccharide chains and most of the synthases are specific to certain sites for coupling reactions to form the glycosidic backbone linkages, despite the multiple hydroxyl and/or amino groups available on the monosaccharide building blocks. The advantages of the specificity of these reactions may be in the control of chain-chain interactions and secondary and higher order structures. Thus molecular recognition events are critical in building complex and responsive systems in biology. For example, the regular β-1-4 linkages between glucose residues in cellulose generate regular secondary structure to provide the appropriate interchain recognition for extensive hydrogen bonding and close packing. This results in strong fibrils essential for plant cell wall structure and mechanical integrity. These systems are driven to higher order via self-assembly, and thus molecular recognition is essential to the successful formation of these complex macromolecular systems. This level of control is a desirable feature that would be useful to be able to predict and utilize in synthetic polymers, yet remains difficult to control using traditional chemical synthesis approaches.

Reactions with enzymes in novel environments and nontraditional solvents have also led to the enhanced ability to control molecular weight and dispersity as well as the morphology and architecture of polymers products. In biology, this control is achieved at the genetic level, since the size the gene determines the size of the transcript and thus the size of the translated protein product. Thus monodisperse protein polymers are the norm in biology based on the mode of synthesis, while monodisperse polymers are difficult to generate with traditional synthetic methods of polymerization. In biology, even with nonprotein-based polymers (e.g., polysaccharides), where direct genetic templates are not utilized to control polymer structure, strong control of the process is still realized. This control is achieved by tight regulation and feedback mechanisms of synthesis pathways, along with modulation of enzyme kinetics, thermodynamics and interfacial activity.

2.2 Ease and Flexibility of Reactions -

An important feature of enzyme reactions in polymer science is the growing range of reaction conditions in which catalysis can be performed. These discoveries have broadened the flexibility of reaction conditions and types of polymer products that can be generated. These advances include alternative solvents such as organic solvents, biphasic organic solvent-aqueous systems, reversed micellar systems and supercritical fluids. These advances have included altering physical and chemical conditions such as temperature, pH, pressure, and salt. This range of reaction environments has significantly expanded from 20 years ago when enzyme reactions were run primarily in aqueous buffered environments within a narrow, near physiological, set of conditions. In addition, one step enzymatic reactions can be considered 'easier' to conduct versus the extensive multistep protection-deprotection schemes often required in synthetic reactions to control polymer structure and reactivity of monomers. For example, the regioselective capabilities of enzymes to form glycosidic linkages contrasts with the multistep blocking and deblocking process used for similar chemical approaches. Stereoselective control of these reactions is an additional benefit in generating 'easier' reactions since high yields of stereochemically pure products is the result. Similar controls with synthetic polymer synthesis often require multistep processes or heavy metal catalysts, and usually result in low yields of stereochemically pure product.

2.3. Green Chemistry -

An increasingly important issue in polymer science is consideration for material life cycles. The use of environmentally compatible processes during synthesis and processing is key to meeting environmental regulations. Enzymes, derived from renewable resources, represent an important option in addressing these needs. Their operation in aqueous systems or in low solvent or solventless systems, and the avoidance of the heavy metal catalysts for control of specificity, are supportive reasons for the use of enzymes in polymer synthesis. Finally, the products of these reactions, whether polyesters, polyphenols, proteins or other polymers, are all biodegradable. Thus, under the appropriate environmental conditions, these polymers can be returned to natural geochemical cycles without the problems associated with nondegradable sythetic polymers that accumulate in landfills or as litter.

3. Sources of Enzymes

Enzyme reactions are being studied from many different perspectives. These studies are examining fundamental aspects of the reactions to understand how to manipulate or control the process, or to overcome limitations in the reactions. Studies to modify enzymes are also underway to explore practical applications for these catalysts in the synthesis or modification of polymers. As a result of these studies, many different experimental approaches are being used or developed to provide insight into enzyme reactions. The various approaches can arguably be divided into four major categories of investigation:

- Enzyme Engineering
- Isolation of Novel Enzymes
- Altering the Reaction Environment
- Altering the Substrate

3.1 Enzyme Engineering -

Until recently, the focus of enzyme engineering had been based on traditional chemical mutation and protein engineering methods. These methods have been continuously refined with structure-activity relationship studies and site directed mutagenesis. However, they require some understanding of the crystalline structure of the enzyme in order to determine where and what to alter in the native primary sequence. These changes will direct alterations in the active site or allosteric locations to modulate environmental interactions. Such rational alterations in an enzyme's native structure also necessitate some level of understanding of the mechanism of catalysis. Even with some understanding of the structure and the mechanism of reaction, experimental insights derived from protein engineering are often empirical. More powerful methods for modifying enzymes are becoming available and include DNA shuffling, combinatorial methods and directed evolution. These approaches can be used to identify new enzymes or enzyme activities. For example, random libraries from different organisms can be constructed and screened to identify clones with the desired enzymatic activity. Once identified, the genes can be sequenced, characterized and modified further by traditional mutagenesis protocols or by further shuffling techniques. These methods, when subsequently complemented with traditional mutagenesis approaches become quite powerful. The major advantage of these newer methods is that little needs to be known about the structure of the enzyme or its mechanism. The main requirement to use these methods is a good screening method for the activity desired. Thus, this massive parallel processing approach to identify 'new' or 'modified' enzymes replaces the more traditional serial processes. These methods offer a more rapid approach to identifying new or modified sequences leading to new enzyme activity or optimization of function in a particular environment (Shao and Arnold, 1996). For example, Moore and Arnold (1996) used a directed evolution approach to select an esterase with enhanced reactivity. The modifications found in the 'modified' esterase which exhibited a significant increase in activity, did not involve amino acids in the active site directly involved in the reaction with the antibiotic substrate. Thus, it would have been difficult to identify these useful changes in the native enzyme using the more traditional approaches of protein engineering.

Aside from these newer methods to select for or modify enzymes, enzyme stability remains a key issue for future industrial needs. Traditional methods of immobilization or entrapment remain important to enhance enzyme stability. However, recent methods aimed at forming enzyme crystals through cross-linking with glutaraldehyde appear to be very effective in stabilizing a wide range of enzymes (Khalaf *et al.*, 1994; Margolin, 1996). Methods to modify enzymes to enhance compatibility with solvent have also been developed. This has been accomplished through surfactant ion pairing (Paradkar and Dordick, 1994a; 1994b). Examples include lipid-coated lipases soluble in organic solvent for the synthesis of di- and tri-glycerides from monoglycerides and aliphatic acids (Okahata and Ijiro, 1988), and grafting of vinyl monomers using a lipase to enhance esterification reactions in chloroform (Ito *et al.*, 1994).

3.2 Isolation of Novel Enzymes -

The search for new enzymes from organisms in a wide range of native habitats has intensified in recent years. Much of this revised interest in microbe 'hunting' arose from the successful isolation of Taq polymerase from *Thermus aquaticus* from hot springs. This discovery has led to the widespread application of the polymerase chain reaction technique in molecular biology. This thermostable enzyme illustrates the value of isolating enzymes from organisms living at environmental extremes. This approach also has been spurred by studies of biology in and around hydrothermal vents in the deep ocean. Enzymes isolated from any 'extremophile' provide direct access to new enzymes with potential function in environments of interest, as well as helping to elucidate fundamental mechanisms employed in the stabilization of proteins at these environmental extremes. These insights have tremendous implications in the study of the underlying mechanisms involved in enzyme stability and enzyme function in novel environments. This will be particularly important in industrial settings and in enhancing the economics of enzyme storage and use. Thermozymes, as one example of extremophile sources of new enzymes, operate at 60°C to 125°C. Stabilization mechanisms include hydrophobic interactions, salt-bridges, and hydrogen bonding, and many other factors which resist loss of native structure and prevent covalent bond degradation (Vieille and Zeikus, 1996). Heat-inactivation by covalent modifications rather than protein unfolding appear to be critical inactivation steps. This has led to substitutions for cysteines and asparagines to enhance the stability of mesophilic enzymes in order to broaden their use at higher temperatures.

Aside from enzymes functioning in environmental extremes, efforts to isolate enzymes from native organisms capable of carrying out specific reaction pathways have continued. One of the more vigorous efforts in recent years has focused on enzymes involved in the biosynthesis and degradation of (polyhydroxyalkanoates) in bacteria (Steinbuchel and Valentin, 1995). These efforts have led to transgenic plant production of these polymers (Nawrath *et al.*, 1994) since the genes encoding these enzymes have been isolated and characterized.

3.3. Altering the Reaction Environment -

Efforts to alter the reaction environment in order to change the nature of the polymer products generated by enzymes has also gained momentum in recent years. This approach was pioneered by Klibanov and coworkers about ten years ago when they reported on the extensive array of enzyme reactions in which equilibria could be reversed by carrying out the reactions in organic solvents. Early studies on the use of enzymes such as chymotrypsin and horseradish peroxidase in solvents date back thirty years (Dastoli *et al.*, 1966; Siegel and Roberts, 1996). However, the detailed study of enzymatic catalysis in monophasic organic solvents is a relatively recent development (Dordick, 1989; Akkara *et al.*, 1991).

Altering solvent composition provides a unique window into control of enzyme selectivity (Carrea *et al.*, 1995). Predictive models to explain some of the differences between physiochemical properties of the solvents and the change in enzyme structure in a specific solvent remain unstatisfactory, while the experimental results of these studies remain convincing. For example, these phenomena have been well established with enzyme reactions with small molecules. Hirose *et al.*, (1992) studied the asymmetric hydrolysis of dihydropyridine diesters and the influence of solvent on stereochemical preference. For polymers, these types of studies have not been explored in as much detail, although it is clear that solvent composition has a profound influence on molecular weight and polydispersity in the case of peroxidase-based free-radical coupling reactions (see Dordick *et al.*, 1989; Ayyagari *et al.*, 1995).

Recent studies in supercritical fluids and in reversed micelles further illustrate the versatility of these reactions. These environments further expand the range of altered reaction environments that can be useful in enzyme-based polymer synthesis. In supercritical anhydrous ether and fluoroform polymer molecular weight and dispersity of polyesters synthesized from lipase-catalyzed transesterification of a (bis(2,2,2-trichloroethyl) adipate and diol (1,4-butanediol) were lower than those from conventional step condensation reactions (Chaudhary et al., 1995). Tailored control of molecular weight and polydispersity was also reported through control of pressure.

3.4. Altering the Substrate -

Traditionally, enzyme kinetics have been studied by examining a range of substrates and determining reaction rates on each of these substrates. This approach resulted in tables of relative activities for different substrates upon which specific enzymes would react. Recent approaches are focusing on selection or 'tailoring' of the substrate structure to direct the enzymatic reaction towards a specific polymer product with targeted specific properties. This approach opens new options in mimicking the natural process of membrane-based reactions found in biological systems. Enzymatic reactions at interfaces provide control of the interfacial organization of monomers and the polymer products, controls the hydrophobicity/hydrophilicity partitioning, and helps to control polymer morphology or architecture. Enzymatic reactions at interfaces have been studied with respect to lipases (Verger, 1997). Interfacial activation and conformational changes in a 'lid' structure that occludes the active site of the enzyme is one aspect of reactivity influenced by solvent composition and physiochemical state of the ester molecules. Another recent example of this interfacial approach is the use of modified phenols and anilines, altered by the incorporation of a hydrophobic alkyl chain, to organize and partition monomers at an air-aqueous interface (Bruno et al., 1995b). Once organized in this fashion, an enzyme such as peroxidase can be used to catalyzed a free-radical polymerization. The benefit of altering the substrate in this approach is to utilize the interfacial reaction to lead to the synthesis of regioselective isotactic conjugated polymers that would be otherwise unattainable by traditional enzymatic free radical reactions or even via traditional synthetic approaches. Recent findings in the study of biological free radical coupling reactions demonstrate that stereoselective coupling reactions between phenoxy radicals also can be assisted by another protein (e.g., chaperone-like), even though this accessory protein has no active center (Davin et al., 1997). This 78 kDa protein apparently reacts with the free-radical intermediate with subsequent stereoselective coupling. These reactions were demonstrated with biologically relevant materials, such as coniferyl alcohol, related to lignan biosynthesis. This finding suggests that enzymatic free-radical polymerizations can be controlled to generate stereoselective as well as isotactic regioselective polymers. This finding has significant implication in control of enzyme-based free-radical polymerization process to generate conjugated polymers for electrooptical applications.

4. Polymers Synthesized Via Enzymatic Processes

• Polyesters
• Polysaccharides and Proteins

- Polyaromatics
- Complex Architectures

The following sections introduce the chapters in the book. The summaries are categorized based on the type of polymer synthesized.

4.1 Polyesters -

The field of enzyme-based polyester synthesis is reviewed in the contribution from **Chaudhary** *et al.* The roles of solvent and substrate activation on molecular weight are emphasized. Divinyl adipate and 1,4-butanediol are used as an example of a well controlled system where competing reactions between synthesis and hydrolysis can be manipulated to gain control of the polymer product (Figure 1). Microbial polyesters, such as the polyhydroxyalkanoates are discussed, as is the role of hydrolases in organic solvents for condensation reactions (A-B,AA-BB). Enzymes studied in these reactions have included porcine pancreatic lipase as well as lipases from *Pseudomonas fluorescens*, *Candida rugosa* and *Rhizomucor miehei*, and other microorganisms. The use of matrix assisted laser desorption/ionization mass spectrometry is discussed as an important analytical tool in the analysis of polymer synthesis, along with more traditional techniques such as Nuclear Magnetic Resonance, gel permeation chromatography and titration. Options to regulate molecular weight are also presented and include control of solvent composition. A key step is the control of water to avoid hydrolysis. Thus the elimination of solvent (bulk polymerization) from some systems and dispersing the enzyme in the reactants directly can result in relatively high molecular weight polymer products compared with more conventional solvent-based approaches. To help in further refining the kinetics of these reactions, a computational model is also presented to address the need for more predictable and directed approaches for these reactions to control of molecular weight.

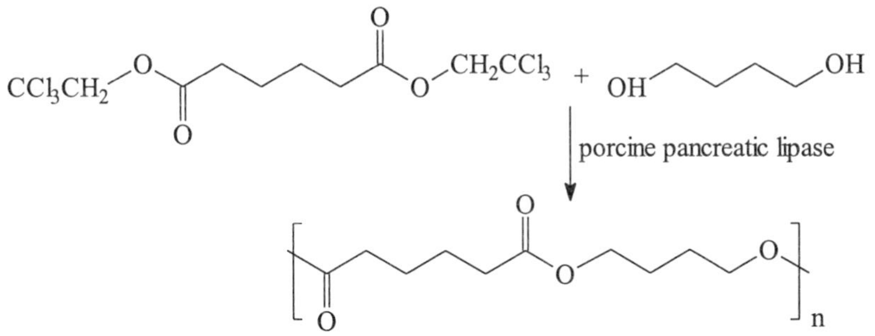

Figure 1. Lipase catalyzed polymerization to form a polyester (Chaudhary *et al.*, 1995).

Polyester synthesis via enzyme-catalyzed ring-opening polymerization is reviewed in the contribution by **Matsumura** *et al.* The main focus of the research is on polymerization of four-membered lactones, β-propiolactone and benzyl-β-malolactonate, using lipase to synthesize poly(β-propiolactone and poly(β-malic acid). Molecular weights up to 50,000 for the poly(β-propiolactone) were found. Porcine pancreatic lipase and other commercial lipases were effective in the reactions which could be run at temperatures up to 60°C. Poly(glycolic acid), poly(malic acid) and poly(β-priopiolactone) have been widely studied and used in biomedical applications as degradable materials. Enzymatic routes to these materials may be advantageous for enhanced control of polymer properties, particularly stereoisomer purity and hydrophobicity (Figure 2). The influence of reaction conditions (time, enzyme concentration) on polymer characteristics is presented. The polymers generated in these reactions were also found to be biodegradable under aerobic and anaerobic conditions.

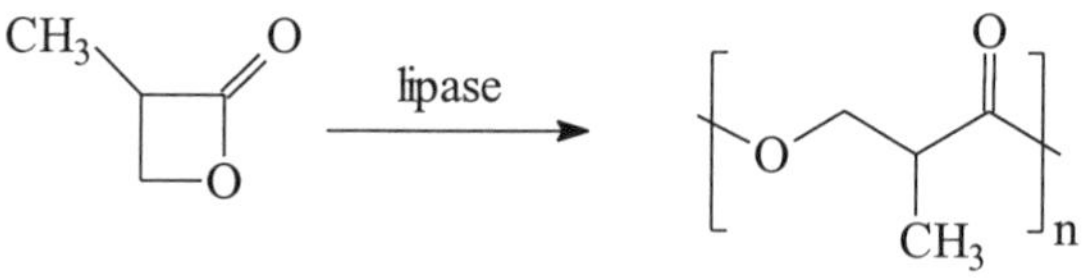

Figure 2. Ring opening polymerization of α-methyl-β-propiolactone by lipase to form a polyester (Svirkin *et al.*, 1996).

Macrolactides appear to be more reactive than smaller sized lactones in lipase reactions in organic solvents to synthesize polyesters. Studies with 11-undecanolide and 15-pentadecanolide resulted in higher reaction rates than the smaller lactones, such as 6-hexanolide (ε-caprolactone) (Uyama *et al.*, 1995a). These reactions can be run without solvent at 60°C. Lipases at 75°C in solvent free systems were effectively used to catalyze the ring-opening polymerization of 12-dodecanolide. Molecular weights above 10,000 were achieved and reaction rates were significantly higher than similar systems with ε-caprolactone as monomer (Uyama *et al.*, 1995a; 1995b). A wide range of novel lipase-catalyzed reactions to synthesize polyesters is addressed in the chapter by **Kobayashi and Uyama**. Ring-opening polymerizations with lactones of different ring sizes (4-membered to 16-membered), methacryl macromonomers with vinyl methacrylate, polycondensation of divinyl adipate with glycols, and succinic anhydride with glycols, to generate aliphatic polyesters were all studied. Lipases from *P. fluorescens*, *Candica rugosa* and porcine pancreatic lipase were used and reaction rates were different depending on the lactone used in the reaction. Correlations between reaction rate and ring strain of the lactone, determined by dipole moments, are discussed. A reaction mechanism based on the formation of a acyl-enzyme intermediate at the serine-active site residue is proposed, with initiation by nucleophilic attack of water on the acyl carbon of the intermediate.

Ring-opening reactions are further addressed in the contribution by **Bisht *et al***. Chemo-enzymatic routes to optically active polymers from racemic monomers is addressed. Specifically, racemic α-methyl- and β-methyl-β-propiolactone) were resolved with lipase (PS30 from *P. fluorescens*) in organic solvent to yield [R]-enriched monomers. The monomers were polymerized to form optically active [R]-poly(α-methyl-β-propiolactone) and [R]-poly(β-methyl-β-propiolactone). Reactions were also carried out with lipase for stereoselective ring-opening polymerization of racemic β-metyl-β-propiolactone) to yield [S]-enriched polymers. Thus, depending on the route chosen, either [R] or [S] enriched polyesters were generated. Porcine pancreatic lipase was used to study the mechanism of ε-caprolactone polymerization. An expression for the rate of propagation was derived from the data and similarities of these reactions to immortal polymerizations are discussed.

4.2 Polysaccharides and Proteins

Cellulose and amylose synthesis *in vivo* proceed via sugar-nucleotide dependent glycosyltransferases, and many of these enzymes are membrane bound. Only recently has significant progress been made toward the synthesis of polysaccharides *in vitro* (Kobaysashi *et al.*, 1991). Cellulose was synthesized from β-cellobiosyl fluoride as the glycosyl donor, with cellulase (Figure 3). The synthesis of oligosaccharides using β-lactosyl fluoride as the glycosyl donor, with regio- and stereo-selectivity in the reactions of a variety of sugars and modified sugars has been reported (Shoda *et al.*, 1993). The lactosylation at the 4-position of the non-reducing end of the glycosyl acceptor provides the site for formation of the (1-4)-β-glycosidic bond. These reactions yield low molecular weight oligomers and reactions can be run in acetate buffer or organic solvent buffer mixtures to optimize synthesis. Xylans can also be synthesized with cellulase and β-xylobiosyl fluoride in acetonitrile acetate buffer (Kobayashi *et al.*, 1996).

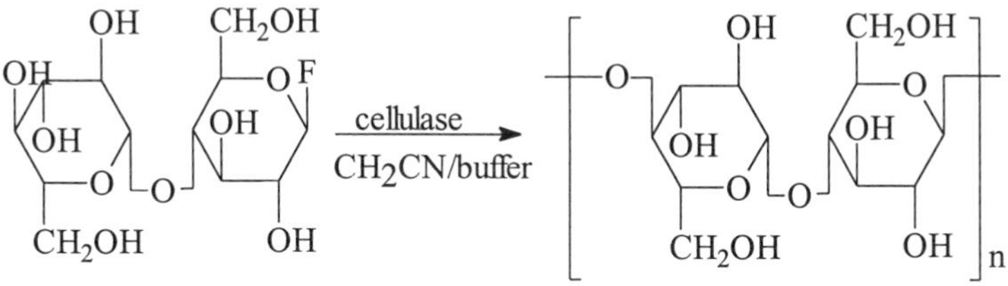

Figure 3. Cellulose synthesis using cellulase (Kobayashi *et al.*, 1991).

Aside from the synthesis of polysaccharides, decoration of polysaccharides with pendant groups is a common activity in industry used to generate a wide range of useful materials. One common example is the synthesis of cellulose acetate from native cellulose. Since polysaccharides are inherently water soluble or swellable, chemical treatments are critical to alter environmental stability and the mechanical properties of the materials. Some of these chemical reactions are difficult to modulate and in many cases extensive protection-deprotection schemes are required to achieve desired structural features. Therefore, solvent-based reactions with surfactant-ion paired enzymes in heterogeneous reactions have been explored to achieve regioselective surface substitutions (e.g., acylations) (Bruno *et al.*, 1995a). The extent and nature of the acylation can be modulated with reaction conditions. Proteases have been used in these reactions. These surface modified polysaccharides may be useful in paper and textile industries to increase hydrophobicity, in food technology for controlling wetting and interfacial properties, as compatibilizers in polymer processing, and in controlled drug release. The reactions are scaleable and continuous production processes are feasible. Only the surface layers are modified by the enzyme reaction while the bulk properties of the polysaccharides remain unchanged. These studies are described in the contribution from **Bruno et al.**

Day et al. discuss a different approach to polysaccharide modification using enzymes. They have developed a regioselective acetylation reaction for alginates using whole cell biocatalysis. Alginates are derived from a wide range of different organisms, and depending on their source include mannuronic acid in acetylated (at the 2-hydroxyl position) and nonacetylated forms. These differences result in variable tendencies to form gels. Using immobilized cells of *Pseudomonas syringae,* continuous acetylation of the 2-hydroxyl position for nonacetylated alginate was demonstrated.

Chitin, a polymer consisting of N-acetylglucosamine monomers linked β-1-4, has been enzymatically modified using tyrosinase (**Lenhart et al.**). Polysaccharide-grafted phenols resulted from these reactions. The oxidation of phenol, catechol, caffeic acid and L-dihydroxyphenylalanine by tyrosinase in the presence of thin films of chitin result in the grafting reactions. The *o*-quinones formed by the enzyme covalently link to the amine groups on the chitin.

Peptide bond formation to generate proteins has been extensively studied and progress has recently been reviewed elsewhere (Sears and Wong, 1996). Two approaches have been used. A thermodynamic approach wherein equilibria are shifted towards peptide bond formation using N-protected amino acids in organic solvents with minimized water content to reverse hydrolysis reactions by mass action. Alternatively, a kinetic approach has been used which involves amino acid or peptide esters as acyl donors in aqueous environments. Here the goal is higher aminolysis rates versus hydrolysis rates.

4.3 Polyaromatics-

Polyphenols can be useful in a wide range of applications depending on their structure. Photoresists, resins, fire-retardant materials, nonlinear optical coatings, biosensors, conductive polymers, and electromagnetic shielding, among others are some examples of possible applications for these polymers. Klibanov and his colleagues first demonstrated that peroxidase in organic solvents catalyzed the synthesis of conjugated polyaromatics through a free-radical coupling reaction (Dordick *et al.*, 1987). Since these studies, detailed structural characterization of the polymers has been reported (Akkara *et al.*, 1991) as well as options to enhance control of polymer molecular weights and dispersities by tailoring solvent composition (Ayyagari *et al.*, 1995). Free-radical polymerization reactions are inherently difficult to modulate. Recent studies demonstrated improve control of polymer structure and morphology by combining enzymatic free radical coupling reactions with interfacial control (Figure 4). By tailoring interfacial reactions and solvent composition, polymer molecular weight, dispersity, regioselectivity, tacticity and architecture can be controlled. Similar interfacial reactions at solvent-aqueous interfaces using reverse micelles enhance control of morphology (spheres), maintain regioselectivity, and enhance control of molecular weight and dispersity by modulation of surfactant, solvent and water phases.

Tailoring solvent composition is also an important route to modulating polyaromatic structure. As an example, with ethanol as solvent, oligomers of *p*-ethyl phenol were synthesized using peroxidase and hydrrogen peroxide with dispersities near one (Ayyagari *et al.*, 1995). In addition, by using solvent mixtures with chloroform, oligomer molecular weights from 1000 to 3000 with low polydispersities were generated. Further expansion of the utility of these polymers through post-polymerization reactions was also demonstrated. For example, the conjugated polymers synthesized from *p*-ethylphenol were functionalized with palmitoyl, cinnamyl and biotin groups since the *p*-hydroxyl groups are not coupled during the free radical polymerization process and thus are available for post-polymerization modification. Polyphenol synthesis using peroxidase reactions is presented by **Ayyagari *et al.*** The main focus of these studies is the interaction between enzyme, polymer and solvent to tailor or control the molecular weight of poly(*m*-cresol). Substrate partitioning between bulk solvent (dimethyl formamide, ethanol, buffer) and the enzyme was assessed through enzyme kinetics. Thermal and structural features of the polymers were determined by Differential Scanning Calorimetry and Fourier Transform Infrared Spectroscopy, respectively

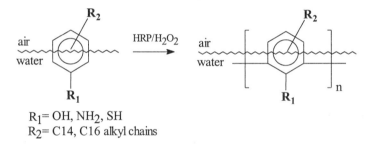

R_1= OH, NH$_2$, SH
R_2= C14, C16 alkyl chains

Figure 4. Peroxidase reactions at an air-aqueous interface carried out on a Langmuir trough to control morphology (two-dimensional thin films), regioselectivity and isotacticity of conjugated polymer products (Bruno *et al.*, 1995b). HRP = horseradish peroxidase.

Peroxidase reactions have been extended when carried out with reversed micelles. These reactions have been used to generate a set of novel polyaromatic compounds in the contribution from **Banerjee et al**. In these studies, the interfacial aspects of aqueous-solvent interfaces are utilized with the aid of a surfactant to control the reaction kinetics and the selectivity in coupling when peroxidase is used with phenolic monomers. Novel microspherical conjugated polymers homogenous in size were generated with sizes in the micron to submicron range depending on reaction conditions. An additional novelty to these reactions is the ability to control the internal density of the microspheres and the encapsulation of enzymes for delivery systems. Luminescent materials were synthesized using similar reactions but starting with naphthols as the monomer. In a similar fashion, when thiolphenols were used as monomers then polymer-seminconductor nanoparticles could be formed.

Kobayashi and Uyama, in their contribution also address the synthesis of polyaromatics using peroxidase catalyzed reactions with phenols and anilines. Polyphenol oxidases were also used to polymerize syringic acid to form poly(phenylene oxide). Most reactions were carried out in dimethyl sulfoxide or dimethyl formamide.

4.4 Complex Architectures

Chemoenzymatic approaches to the synthesis of monomers and polymers is discussed in the chapter from **Ritter**. Enzymes are used to generate methacrylmonomers. Products of the reactions include polyrotaxanes formed from enzymatically derived cyclodextrins that are threaded onto polymers, comb polymers formed from the lipase-catalyzed esterifications of chloic acid, dendrimer formation using peptide chemistry with lysase-catalyzed aspartic acid from fumaric acid, and photosensitive chalcones with peroxidase-based polycondensation reactions with aniline-derivatives.

5. Degradation of Polymers

To fully employ the benefits of enzyme-based technology in polymer science, consideration should be given to material life cycle issues. Therefore, studies into the degradation kinetics of polymers are needed to further understand kinetics and controlling mechanisms. A great deal of study on the degradation mechanisms of polymers, including poly-β-hydroxyalkanoates, polysaccharides, proteins, lignins and related biopolymers has been reported, and no attempt will be made to review these results here. However, in many cases mechanisms are not fully understood. In the contribution by **David et al.**, a detailed study into the hydrolysis of dimethylesters of N-succinyl phenylalanine, a polyesteramide, by papain is presented. Computer simulations are used to elucidate conformations, chain-chain interactions via hydrogen bonding and the potential energies of the substrates in enzyme-complex form. These theoretical data are compared with experimental kinetic data. With advances in computational power and methdologies, this is an important area for future study.

Conclusions

Tremendous new opportunities are becoming available in the use of enzymes in polymer science. The opportunities include enhanced control of enantioselectivity and regioselectivity, the ability to control molecular weight and dispersity, and the ability to synthesize entirely new polymers. With advancements in isolation of enzymes, new techniques for engineering enzymes, the availability of an increasing range of novel reaction environments for enzymes, and better analytical tools to follow enzyme reactions, the field is just beginning to impact traditional synthetic polymer methods. When these advancements are coupled with environmental considerations, the use of enzymes for the formation or modification of polymers is compelling.

References

Akkara, J. A.; K. J. Senecal; D. L. Kaplan. 1991. Synthesis and characterization of polymers produced by horseradish peroxidase in dioxane. J. Polym. Sci. 29:1561.

Ayyagari, M. S.; K. A. Marx; S. K. Tripathy; J. A. Akkara; D. L. Kaplan. 1995. Macromolecules 28:5192.

Bruno, F. F.; J. A. Akkara; M. Ayyagari; D. L. Kaplan; R. Gross; G. Swift; J. S. Dordick. 1995a. Enzymatic modification of insoluble amylose in organic solvents. Macromolecules 28:8881-8883.

Bruno, F. F.; J. A. Akkara; L. A. Samuleson; D. L. Kaplan; B. K. Mandal; K. A. Marx; J. Kumar; S. K. Tripathy. 1995b. Enzymatic mediated synthesis of conjugated polymers at the Langmuir trough air-water interface. Langmuir 11:889-892.

Carrea, G.; G. Ottolina; S. Riva. 1995. Role of solvents in the control of enzyme selectivity in organic media. Trends Biotech. 13:63-70.

Chaudhary, A. K.; E. J. Beckman; A. J. Russell. 1995. Rational control of polymer molecular weight and dispersity during enzyme-catalyzed polyester synthesis in supercritical fluids. J. Am. Chem. Soc. 117:3728-3733.

Davin, L. B.; H.-B. Wang; A. L. Crowell; D. L. Bedgar; D. M. Martin; S. Sarkanen; N. G. Lewis. 1997. Stereoselective biomolecular phenoxy radical coupling by an auxiliary (dirigent) protein without an active center. Science 275:362-366.

Dordick, J. S. 1989. Enzymatic catatlysis in monophasic organic solvents. Enzyme Microb. Technol. 11:194-210.

Dordick, J. S.; M. A. Marletta; A. M. Klibanov. 1987. Polymerization of phenols catalyzed by peroxidase in non-aqueous media. Biotechnol. Bioeng. 30:31-36.

Hirose, Y.; K. Kariya; J. Sasaki; Y. Kurono; H. Ebike; K. Achiwa. 1992. Tetrahdron Lett. 33:7157-7160.

Ito, Y.; H. Fujii; Y. Imanishi. 1994. Modification of lipase with various synthetic polymers and their catalytic activities in organic solvent. Biotechnol. Prog. 10:398-402.

Khalaf, N., C. P. Govardhan, J. J. Lalonde, R. A. Perscihettii, Y.-F. Wang, A. L. Margolin. 1994. JACS 118:5494.

Kobayashi, S.; K. Kashiwa; T. Kawasaki; S.-i. Shoda. 1991. Novel method for polysaccharide synthesis using enzyme: the first in vitro synthesis of cellulose via a nonbiosynthetic path utilizing cellulase as catalyst. J. Am. Chem. Soc. 113:3079-3084.

Kobayashi, S.; X. Wen; S.-i., Shoda. 1996. Specific preparation of artificial xylan: a new approach to polysaccharide synthesis by using cellulase as catalyst. Macromolecules 29:2698-2700.

Koskinen, A. M. P.; A. M. Klibanov. 1996. Enzymatic Reactions in Organic Media. Chapman and Hall, Glasgow, Scotland

Margolin, A. L. 1996. Novel crystalline catalysts. Trends Biotech. 14:223-230.

Moore, J. C.; F. H. Arnold. 1996. Directed evolution of a para-nitrobenzyl esterase for aqueous-organic solvents. Nature Biotechnol. 14:456-467.

Nawrath, C.; Y. Poirier; C. Somerville. 1994. Targeting of the polyhydroxybutyrate biosynthetic pathway to the plastids of *Arabidopsis thaliana* results in high levels of polymer accumulation. Proc. Natl. Acad. Sci. USA 91:12760-12764.

Okahata, Y.; K. Ijiro. 1988. A lipid coated lipase as a new catalyst for triglyceride synthesis in organic solvents. J. Chem. Soc. Chem. Commun. Japan 1392-1394.

Paradkar, V. M.; J. S. Dordick. 1994a. Mechanism of extraction of chymotrypsin into isooctane at very low concentrations of aerosol OT in the absence of reverse micelles. Biotech. Bioeng. 43:529-540.

Paradkar, V. M.; J. S. Dordick. 1994b. Aqueous-like activity of α-chymotrypsin dissolved in nearly anhydrous organic solvents. J. Am. Chem. Soc. 116:5009-5010.

Riva, S.; J. Chopineau; A. p. G. Kieboom; A. M. Klibanov. 1988. Protease-catalyzed regioselective esterification of sugars and related compounds in anhydrous dimethylformamide. J. Am. Chem. Soc. 110:584-589.

Sears, P., C. H. Wong. 1996. Engineering enzymes for bioorganic synthesis: peptide bond formation. Biotech. Progress 12:423-433.

Shao, Z.; F. H. Arnold. 1996. Engineering new functions and altering existing functions. Curr. Opinion Structural Biol. 6:513-518.

Shoda, S.-i., K. Obata; O. Karthaus; S. Kobayashi. 1993. Cellulase-catalyzed, stereoselective synthesis of oligosaccharides. J. Chem. Soc. Chem. Commun. 18:1402-1404

Steinbuchel, A.; H. E. Valentin. 1995. Diversity of bacterial polyhydroxyalkanoic acids. FEMS Microbial. Lett. 128:219-228.

Svirkin, Y. Y.; J. Xu; R. A. Gross; D. L. Kaplan; G. Swift. 1996. Enzyme-catalyzed stereoelective ring-opening polymerization of α-methyl-β-propiolactone. Macromolecules 29:4591-4597.

Uyama, H.; K. Takeya; N. Hoshi; S. Kobayashi. 1995a. Lipase-catalyzed ring-opening polymerization of 12-dodecanolide. Macromol. 28:7046-7050.

Uyama, H.; K. Takeya; S. Kobayashi. 1995b. Enzymatic ring-opening polymerization of lactones to polyesters by lipase catalyst: unusually high reactivity of macrolactides. Bull. Chem. Soc. Jpn. 68:56-61.

Verger, R.1997. 'Interfacial activation' of lipases: facts and artifacts. Trends in Biotechnology 15:32-38.

Vielle, C.; J. G. Zeikus. 1996. Thermozymes: identifying molecular determinants of protein structural and functional stability. Trends Biotech. 14:183-190.

Wang, P. G.; W. Fitz; C-H. Wong. 1995. Making complex carbohydrates via enzymatic routes. Chemtech 4:22-32.

SYNTHESIS

Chapter 2

Enzymes for Polyester Synthesis

Apurva K. Chaudhary, Eric J. Beckman, and Alan J. Russell

Department of Chemical Engineering and Center for Biotechnology and Bioengineering, University of Pittsburgh, Pittsburgh, PA 15261

Abstract : Enzyme-catalyzed AA-BB type polyesterification is a dynamic transesterification system, the outcome of which is determined by substrate structure, stoichiometry, enzyme type and the mode of polymerization. The role of solvent and substrate activation can be pivotal in determining the attainable molecular weight alongwith the strategy used to push reaction equilibrium in the forward direction. For divinyl adipate and 1,4-butanediol system, efficient synthesis of high molecular weight polyester is reported using a commercial lipase. Understanding of how polymerization proceeds provides valuable insight into the role of hydrolysis and the dominant pathways at different stages during the reaction.

Polyesters are heterochain macromolecular compounds containing carboxylate ester groups in the repeating units of their main chains. Varying the molecular architecture and copolymerization conditions, polyesters with wide range of properties *(1,2)* and uses are obtained (Table I).

Polyesters are synthesized by condensation of polyfunctional carboxylic acids (or their derivatives) with polyfunctional alcohols. They are also produced by polycondensation of hydroxycarboxylic acids and ionic polymerization of lactones (cyclic esters) *(3,4)*

In short, the backbone structure, the molecular weight and the end-group functionality together determine the specific application of polyester with end-group functionality critical for applications where further chemical modification is desired. High (>10,000) molecular weight polyesters are an important class of polymers whereas low (< 2,000) molecular weight polyols are used as starting intermediates in synthesis of polyurethane elastomers *(5)*.

Description of Traditional Synthesis of Polyesters

Conditions : Temperature and Pressure. The process of polyester synthesis involves heating hydroxycarboxylic acid or a mixture of a diol and dicarboxylic acid to temperatures at which esterification occurs with the formation of polyester and water as a byproduct. Temperatures in excess of 150 °C, determined by that sufficient for maintaining a homogeneous melt phase reaction and that necessary to prevent thermal degradation during long time period (~ 24 hrs) are used. Since polycondensation is limited by equilibrium, reduced pressures are used to remove water from the reaction mixture. Alternatively, water can also be removed by using an inert carrier gas or an azeotropic distillation.

Catalyst. The self-catalyzed reaction becomes slower with increasing conversion due to the loss of active acid groups. External catalysts are therefore commonly used. These include protonic acids, Lewis acids, titanium alkoxides and dialkyl tin(IV) oxides. Diol dehydration (1,4-butanediol to tetrahydrofuran) and cyclization to form lactones are undesirable side reactions when strong acid catalysts are used. Strongly acidic catalysts also promote discoloration and hydrolysis of the products if they are not neutralized or removed from the polyesters.

Biological Synthesis of Polymers

The use of biological systems for materials synthesis is still a relatively recent endeavor. It is attractive for two reasons. Firstly, using advances in recombinant technology *(6)* it is possible to engineer micro-organisms to synthesize novel or derivatized natural materials (novel polyketides *(7)* with pharmaceutical potential or periodic polypeptides containing non-natural amino acids *(8,9)* and possessing supramolecular architecture *(10)*). Alternatively, enzymes from different sources can be used to perform selective synthesis for generating novel materials. Biologically synthesized polymers will also be readily biodegradable *(11)*.

Microorganisms for Synthesis of Polyesters. Metabolic reactions convert building blocks into large number of natural products. Microorganisms can be isolated or engineered to perform the necessary chemical transformation to produce industrially important monomers and/or polymers. Well-known processes of commercial significance *(6)* of this type are : biological synthesis of derivatized cyclohexadiene followed by thermal aromatization to produce polyphenylene, synthesis of α,ω-dicarboxylic acids, regiospecific hydroxylation of biphenyl to the 4,4'-dihydroxy derivative *(12)* and synthesis of poly(hydroxyalkanoates) *(13)*.

Poly(hydroxyalkanoate) Synthesis. Microorganisms such as bacteria produce biodegradable polyhydroxyalkanoate (PHA) polyesters as an intracellular energy and carbon storage material *(14)* from a variety of different substrates such as sugar, alcohols, n-alkanes, n-alkenes, alkanoic and alkenoic acids. Cell growth and PHA production are dependent on the type of substrate used. The most common PHA, poly(3-hydroxybutyrate) (PHB) has been extensively studied in different bacteria. Under proper conditions, bacteria can accumulate about 75 to 80 % of their dry weight as PHAs. A PHA copolymer (poly(3-hydroxybutyrate-co-3-hydroxyvalerate) developed by ICI's Peter Holmes' group in 1981 is in commercial use in biodegradable films, coatings, molded materials such as bottles and controlled release applications. Zeneca Bio Products marketed this product under the trade name Biopol in United Kingdom. By feeding microorganisms with some other form of carbon source, the type of PHA synthesized can be manipulated and different types of polyesters (that containing 4-hydroxybutyrate units, hydroxyoctanoate units, branched-chain brominated (13) as well as cyanohydroxyalkanoates) have been synthesized.

Enzymatic approach. Enzyme catalysis is inherently more selective than conventional synthesis (15,16). The selectivity can be positional (regioselectivity) or chiral (stereoselectivity). The high selectivity of biocatalysts is useful for reducing side reactions which leads to easier separations. However in addition to the selectivity, enzymes are attractive for commercial applications because of their ability to operate under mild conditions (ambient temperature and pressure). Enzymes are currently available which can catalyze virtually every type of organic synthesis (17). At present almost 3000 enzymes have been recognized by the International Union of Biochemistry (18) and if the speculation that about 25,000 enzymes exist in Nature (19) is true, 90 % of the vast reservoir of biocatalysts still remain to be discovered and used.

Most enzymes function in aqueous environments. In the early 1980s, Prof. Klibanov and coworkers at Massachusetts Institute of Technology demonstrated that enzymes can retain a layer of bound water in organic solvents and thereby maintain catalytic activity. This allows hydrolytic enzymes such as lipases to catalyze polycondensation reactions on hydrophobic substrates in organic solvents. Besides hydrolases, oxidoreductases (horseradish peroxidase and other polyphenol oxidases) form another important class of enzymes investigated for the synthesis of polyphenols (20). Pure enzymatic and chemoenzymatic approaches have been used for the synthesis of wide variety of novel polymers (21,22) for use as water adsorbents, hydrogels, biodegradable materials, chiral resolving matrices, liquid crystals and permselective membranes.

Routes for Enzyme Catalyzed Polyester Synthesis

Hydrolases in organic solvents have been most commonly used for the enzymatic synthesis of polyesters through :

(i) Self-condensation of a hydroxyester or a hydroxyacid (A-B type condensation). In this case, both the reactive functional groups (hydroxyl and acid or ester) are present in the molecule. This type of condensation polymerization is commonly referred to as A-B type polymerization. In general, A-B type monomers have an advantage over A-A and B-B monomers because of the inherent equimolarity of functional groups necessary for obtaining high molecular weight polymer.

Acid-catalyzed A-B type polymerization of hydroxyacid involves competition between intramolecular and intermolecular condensation. The dependence of the lactone-to-oligomer ratio on the chain-length of hydroxyesters/hydroxyacids (based on thermodynamic and kinetic considerations) for acid-catalyzed reactions is well-known *(23)*.

Similar studies have been performed for porcine pancreatic lipase catalyzed polymerization in organic solvents *(24)*. Unsubstituted β, δ and ε-hydroxyesters exclusively undergo intermolecular transesterification to afford corresponding oligomers whereas substituted δ-methyl δ-hydroxyesters undergo lactonization. At lower concentrations and higher temperatures, a higher fraction of product consists of macrocyclic lactones (favorable intramolecular condensation is also observed under similar conditions with acid-catalyzed self-condensation polymerization).

Condensation polymerization of linear ω-hydroxyesters has been investigated in detail by Knani and Kohn *(25)*. They have studied the influence of enzyme type, solvent, concentration, time, size of the reaction mixture and stirring to obtain optimal conditions for the self-condensation of bifunctional methyl 6-hydroxyhexanoate. They observed no enzymatic polytransesterification with aromatic monomers. The degree of polymerization (DP) also decreased with increasing solvent solubility parameter (from 6 to 12 (cal/cc)$^{1/2}$). The DP also followed a S-shaped behaviour with solvent log P (-0.5 < log P < 5) with increase in DP observed around log P ~ 2.5. Decreasing values of DP in good solvents for polyesters were attributed to better solubility and the removal of product oligomers from enzyme surface resulting in reduced substrate concentrations near enzyme. Performing reactions with maximum concentration (in bulk without solvent) afforded polyesters with higher DP (~16 in bulk as compared to ~ 9 in isooctane). They also observed an increase in DP with temperature. In a separate study for enantioselective polymerization of lateral-substituted hydroxyesters, they observed that with increase in the bulkiness of the lateral substituent, in the order Me < Et < Ph, the enzymatic reaction becomes slower while the enantioselectivity is higher *(26)*. Polyesters have been also synthesized by lipase catalyzed ring-opening polymerization of lactones *(25, 27-28)*.

(ii) AA-BB type enzymatic polytransesterification. Since A-B type condensation polymerization involves additional steps for the synthesis and purification of monomer A-B alongwith the possibility for formation of macrocyclic lactones, it is preferable to use AA-BB type condensation polymerization wherein the monomers of the type AA (diacid/diester) and BB (diol) are used. The schematic of such a strategy is represented in Figure 1.

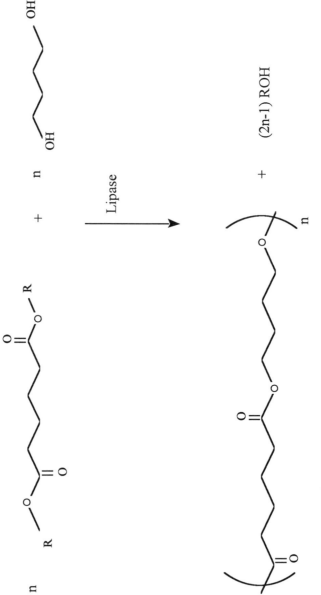

Figure 1. Schematic illustration of AA-BB type condensation polymerization

Design of Substrates and Stoichiometry : Theoretical Considerations.

Choice of Substrates : Binding, Steric and Inductive effects. Enzymatic transesterification in organic solvents follows the acyl-enzyme mechanism. The mechanism consists of the following steps - (i) binding of ester substrate (A) to the enzyme active site with the formation of enzyme-substrate (EA) complex (ii) formation of acyl-enzyme intermediate (EA') with the departure of the alcohol group (P) (iii) the nucleophilic attack of acyl-enzyme intermediate by an alcohol to form transesterified product (Q) and free enzyme (Scheme 1).

Scheme 1. Acyl-Enzyme mechanism for transesterification in organic solvents.

$$E \;+\; A \;\underset{k_{-1}}{\overset{k_1}{\rightleftarrows}}\; EA$$

$$EA \;\underset{k_{-2}}{\overset{k_2}{\rightleftarrows}}\; EA' \;+\; P$$

$$EA' \;+\; B \;\underset{k_{-3}}{\overset{k_3}{\rightleftarrows}}\; E \;+\; Q$$

A = Ester substrate   EA = Enzyme-Substrate (ester) complex
B = Nucleophile substrate   EA'= Acyl Enzyme Intermediate
P = Alcohol product   Q = Ester product
E = Active enzyme (free)

For polyesterification reactions, which are dynamic transesterification systems, the substrates increase in size during the reaction. Therefore, the specificity of the enzyme for starting monomers and oligomers must be considered when designing better substrates for polytransesterification. Polytransesterification is an equilibrium reaction and to obtain high molecular weight polyesters the equilibrium must be pushed in the forward direction. It is therefore important that the alcohol product not be strongly nucleophilic. Using a bulky leaving group in the alcohol moiety should also assist in reducing the rate of backward reaction through steric hinderance.

Choice of Stoichiometry. For a conventional polycondensation reaction, the use of nonstochiometric concentrations of starting monomers will result in the formation of polymers exhibiting those end-groups that are used in excess (23), but the extent of polymerization will also be reduced by stoichiometries other than unity. If the reactive functional groups are utilized exclusively for condensation, then the number average degree of polymerization (X_n) achieved will be sensitive to the nonstoichiometry of starting monomers as predicted by Carothers equation

$$X_n = \frac{1+r}{1+r-2rp} \qquad (1)$$

where r is the ratio of functional groups (< 1) used initially and p is the extent of reaction.

Experimental Deductions. During initial attempts of the enzymatic synthesis of polyesters by AA-BB type condensation, starting stoichiometry of substrates was not a focus. Okumara and coworkers *(29)* were the first to attempt the synthesis of oligoesters using dicarboxylic acids (BB) and diols (AA) as monomer substrates. They performed reactions in a diol/water biphasic system. The only products examined in detail were "trimer", "pentamer" and "heptamer" of the forms AA-BB-AA, AA-BB-AA-BB-AA and AA-BB-AA-BB-AA-BB-AA, respectively. Due to the low solubility of higher oligomers in the reaction mixture and since a large excess (80 times) of diol was used, very low degrees of polymerization and poor yields of oligomers were obtained.

Margolin and coworkers *(30)* described porcine pancreatic lipase catalyzed stereoselective oligomerization for synthesis of optically active polymers. They described condensation between (a) racemic diester (BB) and a diol (AA) (bis(2-chloroethyl) ± 2,5-dibromoadipate with 1,6-hexanediol and bis(2,2,2-trichloroethyl) (±)-3-methyladipate with 1,6-hexanediol) and that of (b) racemic diol with diester (bis(2-chloroethyl) adipate with (±)-2,4-pentanediol). Based on IR and ^{1}H-NMR analysis of the products, they concluded that trimer and pentamer of the type AA-BB-AA and AA-BB-AA-BB-AA and very low quantity of higher oligomers were formed in both the cases. Formation of hydroxyl capped oligomers in the first case (a) is expected because the diol is used in excess, however for the latter case (b) formation of hydroxyl capped oligomers was very surprising because only half of the racemic diol mixture was active in transesterification.

Morrow and colleagues realized the importance of stoichiometry and studied polytransesterification using equimolar quantities of starting monomers (trihaloalkyl diesters and primary diols). Based on the substrate design considerations mentioned above, the trihaloalkyl group was expected to increase the electrophilicity of the acyl carbonyl and secondary diols were avoided for possible rate reduction by steric hinderance *(31,32)*. Further they observed an increase in the rate and extent of polymerization in going from trichloroethyl to trifluoroethyl group. A similar increase was also expected when chlorophenyl moiety was selected as a leaving group. However, this was not observed implying that besides increasing the electronegativity of the leaving group, an enzyme's ability to attack the ester carbonyl is also an important consideration. No polymerization was observed with aromatic diesters and it was suggested that more studies may be necessary to delineate the role of solvent.

Morrow also investigated a stereoselective oligomerization *(33)* between a racemic diester (bis (2,2,2-trichloroethyl) (±)-3,4-epoxyadipate and 1,4-butanediol. The optical purity of the synthesized polymer was estimated to be 95 %.

Linko and colleagues studied polymerization of bis(2,2,2-trifluoroethyl) sebacate and aliphatic diols using M. miehei lipase *(34)*. Identical observations were also made for the effect of diol chain length with the sebacate diester. Similar observations were also made by Uyama and coworkers and Athavale and colleagues for the effect of diol chain length.

Using enol esters as acylating agents will result in the byproduct alcohol which tautomerizes to a ketone or an aldehyde *(35)*. Enol esters not only accelerate the rate of acyl transfer but also shift the reaction equilibrium continuously. Oxime esters are also irreversible acyl transfer agents for lipase catalysis since the byproduct alcohol is a very weak nucleophile and is unable to compete with the substrate alcohol for acyl-enzyme intermediate *(36)*. However, co-substrate inhibition and problems in separating the non-volatile oxime from the substrate alcohol during work-up may be encountered *(17)*.

Limited efforts have been made in past using the above irreversible approach for lipase catalyzed polyester synthesis. Uyama and colleagues *(37)* used divinyl adipate as diester substrate (lipase P from *Psuedomonas fluorescens*) and Athavale and colleagues*(38)* used bis(2,3-butane dione monoxime) alkanedioates as diester substrate (Lipozyme IM-20 from Novo Nordisk) with different alcohols. For the case where divinyl adipate was used, about 98 % of the product with molecular weight of 1500 was obtained after 48 hours before using the reprecipitation/purification technique to isolate high molecular weight fraction. No attempts were made to characterize the end-groups of the polyester (the backbone structure of polyester was however confirmed by NMR).

We have investigated the use of divinyl adipate (DVA) as an activated diester for our polymerization reactions. Acyl transfer using enol esters has been shown to be about 10 times slower than hydrolysis and about 10-100 times faster than acyl transfer using 'activated' esters *(35)*. (For comparison, non-activated esters such as ethyl acetate, reaction rates of about 10^{-3} to 10^{-4} of that of hydrolytic reaction are observed). The vinyl functional group not only activates the ester end-group for faster transesterification but also imparts inherent irreversibility to the process for continuous shifting of reaction equilibrium.

We performed polytransesterifications at different initial ratios of DVA to 1,4-butanediol concentrations analyzing both the initial removal of monomers and the properties of the final product. The value of p obtained with an equimolar

Table I. Polyesters : Method of Synthesis and Applications

Polyester nature	Synthesis Strategy	Application
High molecular weight linear polyesters (M > 10,000)	a) Dicarboxylic acids (or their diester or dichloride derivatives) with difunctional alcohols b) From lactones	Fibers,films, thermoplastics Examples: poly(butylene terephthalate) poly(ethylene terephthalate) e.g. ε-caprolactone
Low Molecular weight (linear or slightly branched) polyesters	a) Saturated aliphatic or aromatic diacids with difunctional or mixtures of di- and trifunctional alcohols b) Di/tri/polyfunctional alcohols with poly-functional usually aromatic dicarboxylic acids or saturated or unsaturated fatty acids	M < 5000 and with hydroxyl end groups at both ends are used as *macrodiols* for synthesis of elastomeric polyurethanes. Plasticizers, alkyd resins. Unsaturated polyesters can copolymerize with monomers such as styrene to form thermosets. Polyester resins
High temperature resistant polyarylates	Polyesters of dihydric phenols with aromatic dicarboxylic acids	Liquid crystalline melts processable to solid products of unusually high mechanical properties.

Table II. Effect of substrate stoichiometry on degree of polymerization for Novozym catalyzed polytransesterification between divinyl adipate and 1,4-butanediol.

DVA/1,4-BD ratio	r	Predicted number average degree of polymerization[a]	Number average degree of polymerization obtained
1.0	1.0	-	24
1.15	0.87	9.2	8.9
1.23	0.813	7.1	7.8
1.5	0.67	4.3	4.8

[a] Using extent of reaction (p) = 0.958 (from r=1.0) for predicting X_n for non-stoichiometric concentration of substrates.

SOURCE: Reprinted with permission from Chaudhary et al. *Biotechnol.Bioeng,*
Copyright 1996 Wiley-Liss, Inc., a subsidiary of John Wiley & Sons, Inc.

mixture of substrates (p=0.958) is used for the determination of X_n for the nonstoichiometric ratio experiments. Experimentally determined values are comparable to the predicted values (Table II). For the biocatalytic step-polymerization reaction between difunctional monomers, non-stoichiometric concentrations of substrates clearly reduce the achievable degree of polymerization.

Choice of Enzyme.

Most of the research in this field was performed using a crude enzyme preparation of porcine pancreatic lipase (PPL). Lately, some attempts have been made using lipase from *P. fluorescens, Candida rugosa* (CRL), *Rhizomucor miehei* (MRL) and better rates of transesterification have been observed with these enzymes. Selection of a suitable enzyme can play a significant role in polyester synthesis process. It is interesting to note that the enzymes (In the increasing order CRL, PPL and MRL) that worked well for polyester synthesis process were also quite active for hydrolytic asymmetrization of *meso*-bis(acyloxymethyl)tetrahydrofurans (39). The substrate for the above reaction has two butanoate ester groups in the backbone. Recently, it has been also postulated that the open cleft on the surface of protein for R. miehei lipase may be favorable for polyester synthesis *(40)*.

We have also previously reported the use of crude porcine pancreatic lipase for the polytransesterification reaction after an initial screening procedure *(21)*. Crude enzyme is most effective for polyester synthesis however, very large quantities of enzyme are required. Previously, effective resolution of aminoalcohol *(41)* precursors by esterification with capronic acid or transesterification with vinyl acetate has been reported using Novozym-435 (an immobilized preparation of B-lipase derived from Candida antarctica). After a suitable screening procedure with divinyl adipate as substrate, we have determined that Novozym-435 is the most effective enzyme.

Method of Polyester Analysis used and Limitations (Determination of Molecular Weight and End-group Functionality).

A variety of techniques have been used previously to analyze the molecular weight and functionality of enzyme-synthesized polyesters. Most important among them are GPC, ^{1}H-NMR, Mass Spectrometry and direct acid titration. In this section we will outline the advantages and disadvantages of different techniques for the analysis of polyesters.

Gel Permeation Chromatography (GPC). GPC is a size exclusion chromatographic (SEC) technique using columns with varying pore sizes to separate polymer chains on the basis of their size in solution. Since monodisperse standards of polyesters are currently not available commercially, GPC analysis of polyesters is usually performed by comparison against standards of polystyrene or poly(ethylene glycol). Typically, a calibration curve of log [Molecular weight] against retention time is obtained for a given set of columns. Molecular weight of the samples is then calculated based on the elution time of the polyester. Molecular weight averages can be overestimated if the obtained values are not compared with absolute values, which can be determined using different other techniques such as light scattering, mass spectrometry etc.

Most importantly, since the molecular weight calibration is sensitive to retention time on the column (a semi-log plot of log [M] vs retention time), large errors in polymer molecular weight estimation can be introduced if (a) either the separation is poor (that is the columns used are not effective for the molecular weight range within which separation of polymer is desired or (b) Too few calibration standards differing from each other considerably in molecular weight are used to construct calibration. This is indeed the case as we have observed previously by comparison of the obtained GPC calculated values using mass spectrometry as well as by using a universal calibration curve of log [M x η] vs retention time *(42)* (Figures 2a and 2b).

[1]**H-NMR.** Attempts were made by different research groups to determine polyester number average molecular weights by end-group analysis using ^{1}H-NMR. With increasing molecular weight, ^{1}H-NMR loses its accuracy because the absorption of endgroups becomes too small to enable accurate integration and calculations. Since then there has been development in the understanding of how enzymatic polymerization works and it has been established that hydrolysis is a significant side reaction during polytransesterification. This adds to the limitation of the proton-NMR technique for the calculation of average molecular weight.

Mass Spectrometry. Mass Spectrometry analysis of synthetic polymers has received considerable attention recently. To be able to accurately determine the distribution of chains and molecular weight averages, it is crucial to avoid fragmentation during desorption and ionization processes involved during mass spectrometry. Matrix assisted laser desorption/ionization (MALDI), has been successfully used for the characterization of a variety of synthetic polymers *(43,44)* and polypeptides *(45)*. The advances in the instrumentation (advent of delayed extraction source) have further made possible the analysis of polymers of higher molecular weights (Critchley, G., Micromass Inc., UK, personal communication, 1996).

MALDI-MS has without doubt been very valuable for structural characterization of individual oligomers *(46,47)*. Considerable attention has been focused on the ability of MALDI to provide accurate molecular weight distributions *(48,49)*. Mass spectrometry has the potential to provide absolute molecular weight distributions that are more accurate than those that are determined by size exclusion chromatography (SEC), light scattering and osmometry. However, since synthetic polymers are a mixture of oligomers, accurate molecular weight analysis requires that both the mass and the abundance of each oligomer species be correctly measured

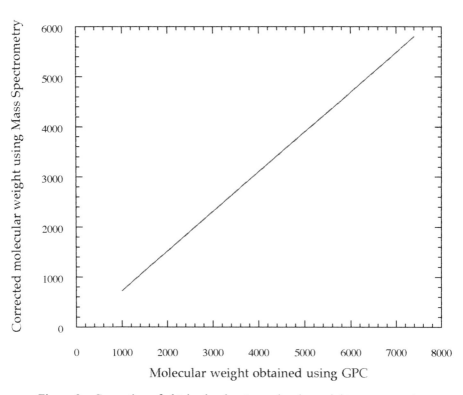

Figure 2a. Correction of obtained polyester molecular weight averages using Mass spectrometry (Adapted from Ref. 42).

continued on next page

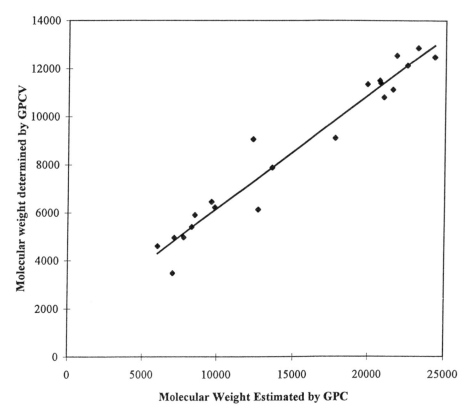

Figure 2b. Comparison of polyester molecular weight averages obtained using GPC with GPCV.

over a wide mass range. An accurate abundance measurements for polymers requires that ionization process, ion transmission and detection be independent of mass (or a known function of mass) and oligomer structure (end-groups, number of repeat units etc.)

MALDI is an ionization method which itself has considerable variation, especially in sample preparations as well as instrumental effects. Variations in time of flight mass analyzers (because of differences in ion acceleration, ion focusing, ion detection etc.) can further cause differences in molecular weight distributions. It is therefore believed that instrumentation or sample preparation are responsible for differences observed with polymer molecular weight distributions *(49)* using MALDI-MS. Several guidelines have been therefore suggested to overcome these limitations :

1. Use of a properly designed instrument and a suitable matrix which will provide accurate mass analysis for a given polymer type.
2. Using a high voltage provides ions with velocity sufficient to impact the detector to eliminate detector mass discrimination.

Using MALDI-MS technique along with SEC can therefore be a powerful apparatus for following the progress of polymerization (Chaudhary et. al., *Biotech. Bioengg.,* submitted). The gradual increase in polymer molecular weight can be easily and accurately monitored. Moreover the changes in the polymerization mixture can be followed with respect to the size as well as the type of chains (end-groups).

Direct Titration. Since hydrolysis occurs during polyesterification, direct titration of end-groups can be used to determine the overall acid content of the polymer. This procedure requires large amounts of sample (hundreds of grams of sample) if the acid content of the polyester is very low (< 5 %). Determination of trace amounts of acid-terminated polyester species is virtually impossible using this method.

Role of Solvent

For any enzyme-catalyzed process in organic solvents, solvent plays a crucial role in terms of altering enzyme activity and stability. For a reaction involving different substrates, there is an added effect of varying enzyme specificity in different organic solvents. For polytransesterification, besides these effects, the solubility of growing polymer chains in reaction solvent and the ability to be able to drive the reaction in the forward direction has to a large extent determined the choice of the solvent.

Morrow and coworkers observed that PPL catalyzed transesterifications were optimal in ether solvents such as diethyl ether, THF and bis(2-ethoxyethyl) ether (2-EEE). They observed that both THF and 2-EEE support detrimental hydrolysis reactions apparently because the high solubility of water in these solvents favors the release of protein bound water. Aromatic solvents such as ortho-dichlorobenzene and dibenzyl ether do not support hydrolysis reactions but the transesterification rate is slow. Selection of solvents in Morrow's case was however determined by the solvent boiling point because they wanted to remove the condensation byproduct trifluoroethanol by evaporation under vacuum. Similar considerations were important in work performed by Linko and colleagues and they used diphenyl ether as solvent. Athavale and coworkers using bis(2,3-butane dione monoxime) alkanedioates and diols, obtained reduced molecular weights in polar solvents which can sustain polymerization and observed polymer precipitation in nonpolar solvents. Uyama and Kobayashi have performed enzymatic polymerization using divinyl adipate with various glycols and observed no increase in the molecular weight of the polyester once the polyester precipitated in diethyl ether with the increase in yield of polyester to 50 % during this period.

Geresh and Gilboa have also attempted the synthesis of "all trans" alkyd resins using lipase catalyzed transesterification of unsaturated diesters with diols *(50,51)*. They observed termination of polymerization at lower molecular weights (600-800) when the reaction was performed in tetrahydrofuran. However for the same reaction in acetonitrile, all-trans polyester which precipitated from reaction had a sharp melting point of 103 °C. The reaction could be taken to completion by filtering the solid material off from the reaction vessel from time to time. They suggested that while chemical synthesis involves combination of oligomers formed initially, enzymatic synthesis involves a regular and controlled addition mechanism *(52)*.

We have previously also reported porcine pancreatic lipase catalyzed polyester synthesis in supercritical fluoroform. Due to the tunable nature of supercritical solvent environment, the reaction was terminated by precipitation of the growing chains by pressure adjustment. Moreover, the precipitated fractions were also analyzed to be essentially monodisperse. We have also investigated the role of solvent for divinyl adipate catalyzed polytransesterification in detail using solution as well as bulk polymerization.

Evolution of molecular weight and end-group analysis

Heterogeneous enzyme-catalyzed condensation involves an additional step for binding of substrate as compared to homogeneously catalyzed chemical condensation. This can therefore lead to specificity differences with the enzyme. Attempts have been made previously to follow the polymerization process. Morrow and coworkers observed that whereas 1,4-butanediol completely disappears from the reaction within the first 6 hours, 20 % of the starting bis(trichloroethyl) ester still remains indicating that the dimer immediately reacts with the diol to form a hydroxyl capped trimer *(31)*. The presence of trichloroethyl ester and hydroxymethyl groups led them to conclude initially that hydrolysis is not responsible for the loss of reactive functional groups and low degrees of polymerization.

In a relatively recent publication, Morrow and coworkers analyzed polymerization growth between bis(trifluoroethyl) ester and 1,4-butanediol using ^{1}H-NMR as well as GPC. They observed an order of magnitude increase in polymer molecular weight by ^{1}H-NMR observation (from 4600 to 37000), which was not the case when they studied the reaction by GPC *(53)*. Unequal areas for absorption of trifluoroethyl ester and hydroxymethyl groups were also observed. Although only hydroxyl and trihaloalkyl end-groups were detected via ^{1}H NMR, they hypothesized that lower number of trifluoroethyl end-groups were a result of their loss by hydrolysis. They observed higher loss of bis(trifluoroethyl) ester groups compared to 1,4-butanediol. They hypothesized the differences in their observations using ^{1}H-NMR and GPC were the result of hydrolysis of active trifluoroethyl groups.

Unfortunately, it is difficult to detect carboxylic acid protons using ^{1}H NMR unless the acid termini are first converted to ester derivatives *(28)*. Insight into how enzymatic polymerization proceeds can also be obtained by comparison of AA-BB type polymerization with an identical A-B type polymerization. Observations made by Morrow showed larger DP in the AA-BB case and no explanation was offered *(32)*.

For the divinyl adipate-1,4-butanediol system, we have systematically followed the progress of polymerization in THF as solvent using GPC and MALDI-MS.

Disappearance of Monomers. A large number of the biocatalytically synthesized polyesters that we have analyzed by ^{1}H-NMR show the presence of -OH functional groups and no termination with vinyl (-CH=CH$_2$) end-groups. Although previous work has given similar results *(37)*, this result was particularly surprising since monitoring of the initial loss of monomers from the reaction mixture demonstrated that the divinyl adipate monomer was reacting faster than the butanediol (Figure 3). Possible side reactions which could account for the non-stoichiometric disappearance of the vinyl group include binding of the monomer to the solid enzyme, homopolymerization of vinyl end-capped oligomers, or the hydrolysis of vinyl adipate to acids. During dimerization, nonstoichiometry cannot be caused by enzyme specificity differences except if DVA reacts faster with a dimer than butanediol, this can result in non-stoichiometric removal of DVA relative to 1,4-butanediol (Figure 4).

Based on the results, we have included a network of reactions occuring during an enzyme-catalyzed AA-BB type polytransesterification (Figure 5). We have also attempted to determine the order and importance of these reactions as the polymerization proceeds.

End group analysis of Polyesters. End-group analysis of polyesters is particularly important for their use as polyols in the urethane industry. Researchers have

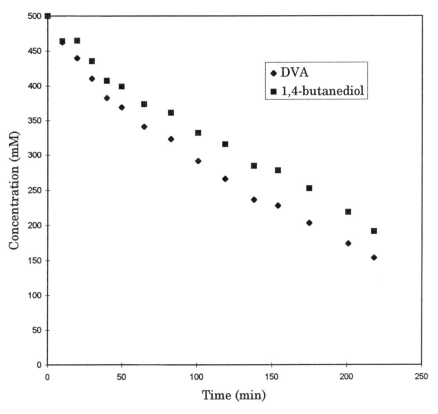

Figure 3. The disappearance of monomers during initial phase of lipase catalyzed polytransesterification reaction between divinyl adipate and 1,4-butanediol (THF as solvent, [DVA] = 500 mM, [1,4-BD] = 500 mM, Temp = 30 °C, shaking speed = 250 rpm).
SOURCE: Reprinted   with   permission   from   Chaudhary   et   al.
Biotechnol.Bioeng, Copyright 1996 Wiley-Liss, Inc., a subsidiary of John Wiley & Sons, Inc.

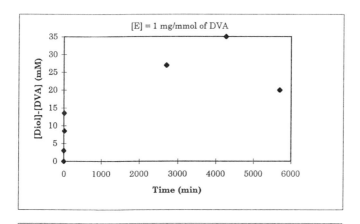

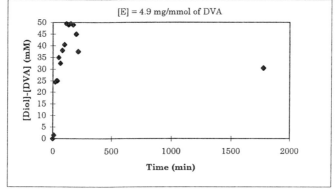

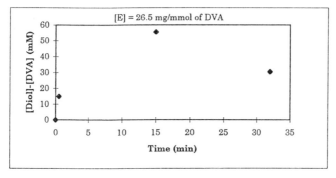

Figure 4. Effect of enzyme concentration on the excess of 1,4-butanediol observed during reaction with divinyl adipate ([DVA]=[1,4-BD]=500 mM, Temp=30 °C, shaking speed=250 rpm)

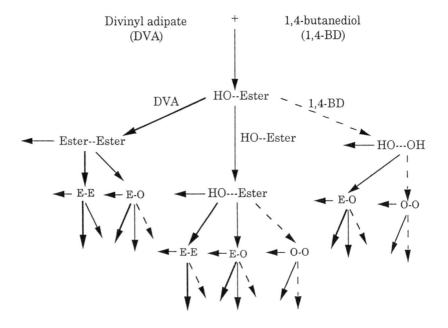

E-E = ester-ester capped species
E-O = ester-hydroxyl capped species
O-O = hydroxyl-hydroxyl capped species

Arrows pointing to the left indicate hydrolysis side reaction

Figure 5. The role of end-group hydrolysis in enzymatic polytransesterification reaction.
SOURCE: Reprinted with permission from Chaudhary et al. *Biotechnol.Bioeng,* Copyright 1996 Wiley-Liss, Inc., a subsidiary of John Wiley & Sons, Inc.

hypothesized that degradation reactions observed during the later stages of polymerization are hydrolytic in nature with the water obtained from the enzyme. We have used MALDI-MS and a direct titration method *(54)* and characterized the nature and extent of acid-end groups within the polyester caused by the onset of hydrolysis of oligomers. MALDI-MS analysis of polyesters after long reaction times revealed the presence of acid-ol, ol-ol and acid-ol species. Experiments were also performed for shorter times and lower enzyme concentrations, in which case ester end-groups of the polyester were detected.

Most importantly though, MALDI-MS has enabled us to investigate the kinetics of polymerization reaction and provide valuable information on the significant processes during different stages of polymerization which are discussed below.

Stages of Enzymatic Polymerization. Since the changes occurring towards the end of polymerization are important in terminating the growth of polymer, the first experiments performed were aimed to observe changes occurring during this phase. MALDI-MS analysis of polyesters grown at different times was analyzed after the point when GPC analysis did not indicate polymerization growth.

Figure 6 shows that polyester which had "grown" for 6 hours (7 % of the species within MW range > 2000), subsequently depolymerizes at the 24 hour stage (less than 1.8 % of species within MW range > 2000, with the complete absence of high mass range species (> 2450) from the polyester) if the enzyme was not removed from the reaction mixture. There is also no apparent change in the molecular weight distribution (the size and the type of polyester) from 36 to 72 hours of reaction. Therefore, after 36 hours, equilibrium has been attained, or the enzyme has lost activity (Figure 6).

After 6 hours, only 3.6 % of the low molecular weight ester-ester terminated species were detected, implying that ester-hydrolysis on oligomers is already occuring. In order to observe the intermediate stage of polymerization, where ester-ester groups should be common, we performed experiments at lower enzyme concentration (Figures 7 and 8). Perhaps the most interesting observation is the absence of higher ester-ester terminated species in MW range > 1600. Within this range, only acid-acid, acid-ol and ol-ol species are detected. This therefore implies that end-group hydrolysis of higher molecular weight polyester species precedes that of the lower molecular weight ester-ester terminated species. This therefore implies that hydrolytic specificity of the lipase used is greater for larger oligomers.

Based on the above observations, we have compiled the chronolgical sequence of events during enzymatic polytransesterification in Figure 9. The relative contribution and the onset of different processes depicted in Figure 8 will be strongly dependent on enzyme water content, enzyme/substrate ratio, substrate stoichiometry, temperature and the type of solvent. Manipulation of either the proportion, or the

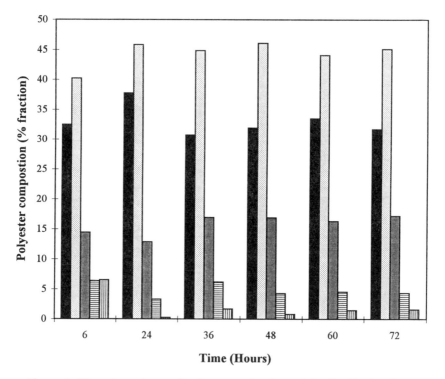

Figure 6. Time dependence of polymer molecular weight distribution for a lipase catalyzed polytransesterification. The use of very high enzyme concentrations ensures that these changes are end-stage alterations. ([DVA] = [1,4-BD] = 500 mM, [E] = 300 mg/mol of DVA]
SOURCE: Reprinted with permission from Chaudhary et al. *Biotechnol.Bioeng,* Copyright 1996 Wiley-Liss, Inc., a subsidiary of John Wiley & Sons, Inc.

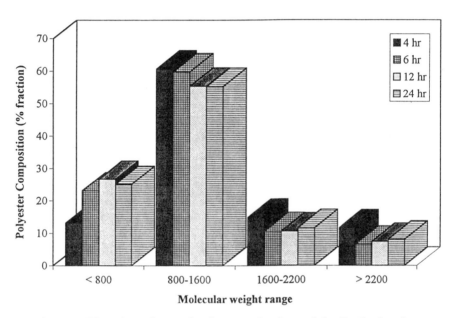

Figure 7. Time dependence of polymer molecular weight distribution for a lipase catalyzed polytransesterification at low enzyme concentrations. ([DVA] = [1,4-BD] = 500 mM, [E] = 20 mg/mmol of DVA]
SOURCE: Reprinted with permission from Chaudhary et al. *Biotechnol.Bioeng,* Copyright 1996 Wiley-Liss, Inc., a subsidiary of John Wiley & Sons, Inc.

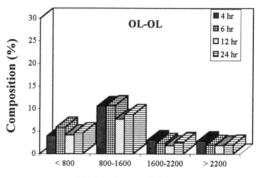

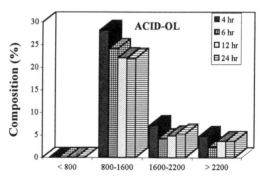

Figure 8. Time dependence of polymer molecular weight distribution for various classes of oligomer during lipase catalyzed polytransesterification at low enzyme concentrations. ([DVA] = [1,4-BD] = 500 mM, [E] = 20 mg/mmol of DVA]

SOURCE: Reprinted with permission from Chaudhary et al. *Biotechnol.Bioeng*, Copyright 1996 Wiley-Liss, Inc., a subsidiary of John Wiley & Sons, Inc.

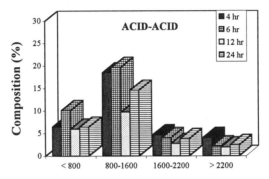

Molecular weight range

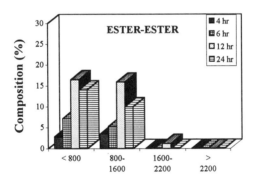

Molecular weight range

Figure 8. *Continued.*

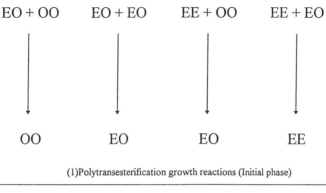

(1)Polytransesterification growth reactions (Initial phase)

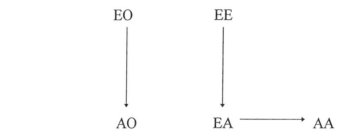

(2) End-group hydrolysis reactions (Intermediate phase)

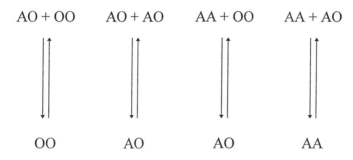

(3) Condensation between acid and alcohol groups and hydrolysis of ester groups
within polyester backbone.

Figure 9. Summary of all possible processes during enzyme-catalyzed polyester synthesis. where E- represents a single ester terminus in an oligomer, A- a single acid terminus and O- a single alcohol terminus. Hydrolysis reactions of backbone esters of EE and EO species are not included.

SOURCE: Reprinted with permission from Chaudhary et al. *Biotechnol.Bioeng,* Copyright 1996 Wiley-Liss, Inc., a subsidiary of John Wiley & Sons, Inc.

onset time of these processes can be used to control polymer molecular weight and functionality.

Efforts to increase polymer molecular weight

Efforts were made in 1992 by Morrow to obtain high molecular weight polyesters by enzymatic polytransesterification. Morrow suggested that due to the hydrophilic nature of protein matrix, and despite drying over phosphorus pentoxide for several days, hydrolysis of active ester end-groups is possible *(55)*. Release of protein bound water appeared more likely in a solvent in the presence of alcohol and a solvent in which water had some solubility. Since a large amount of trifluoroethanol was present during the late stages of polymerization, hydrolysis could be significant during the end-stages of polymerization. This problem therefore led Morrow to select solvents which had relatively high boiling points and use vacuum evaporation to remove byproduct alcohol periodically to obtain high molecular weight polyesters.

Linko and colleagues have used identical methodology (continuous removal of byproduct alcohol using high vacuum and high boiling solvents) with bis(2,2,2-trifluoroethyl) sebacate and aliphatic diols with M. miehei lipase to obtain high molecular weight polyesters *(34)*. They observed that increasing substrate concentration helped increase polymer molecular weight, however increased viscosity with higher concentrations led to poor mixing.

Uyama and coworkers obtained reported molecular weights of polymer of 6700 at 45 °C in diisopropyl ether with 1,4-butanediol after 48 hrs of reaction and purification by reprecipitation. Binns and coworkers*(56)* studied polyesterification using an adipic acid/1,4-butanediol system. They used 4 A° molecular sieves to remove water and midpoint aqueous extraction to improve molecular weight by shifting equilibrium and extracting lower oligomers. To summarize, we present here a compilation of highest achievable polyester molecular weights by enzyme catalysis (Table III).

In our attempts to obtain high molecular weight polyester, we have analyzed directly the product of the reaction without removing low oligomeric species. (Note that removal of low oligomeric species by partial fractionation/precipitation procedure can show high molecular weight and low dispersity at the expense of reaction yield).

Table III demonstrates that high concentrations of enzyme, a viable method for removal of condensation byproduct, and long time periods must be used for the synthesis of even low molecular weight polyesters.

From a purely economic standpoint, if the enzymatic approach has to be competitive with the chemical synthesis (the price of chemically synthesized

Table III. Biocatalytic polytransesterification for polyester synthesis.

Enzyme (Concentration)	Substrate (Solvent) and details	Reaction Time	Molecular weight (Analysis Details)	Ref.
A. niger lipase.	1,13-tridecanedioic acid with 1,3-Propanediol (water-diol mixture)	16 hrs	Trimer, Pentamer and Heptamer by GPC analysis against PEG standards.	(29)
A. niger lipase (4600 g/mol of diester)	bis(2-chloroethyl) (±)-2,5-bromoadipate with 1,6-hexanediol (toluene).	7 days	Trimer and Pentamer by GPC Analysis against PEG standards	(30)
porcine pancreatic lipase (570 g/mol of diol)	bis(2,2,2-trichloroethyl) trans-3-hexanedioate (racemic mixture) with 1,4-butanediol (ether).	3.5 days	M_w=7,900 (after ether fractionation) by GPC analysis against PS standards.	(31)
porcine pancreatic lipase (250 g/mol of diester)	bis(2,2,2-trifluoroethyl) glutarate with 1,4-butanediol (1,2 dimethoxy-benzene). Continuous removal of byproduct trifluoroethanol by evaporation under vacuum.	432 hrs	M_w=39,000 by GPC against polystyrene standards with one ultrastyragel column.	(55)
P. fluorescens (lipase P) (500 g/mol of diester)	divinyl adipate with 1,4-butanediol (di-isopropyl ether)	48 hrs	M_n=6,700 (methanol insoluble fraction) by GPC (details of analysis not available)	(37)
Mucor Miehei lipase (100 g/mol of diester)	bis(2,3-butane dione monoxime) glutarate with 1,6-hexanediol in di-isopropyl ether.	48 hrs	M_w=9,200 by GPC. (Analysis against 4 PS standards)	(38)
Lipases from Pseudomonas and Mucor (amount unknown)	Dichloroethyl fumarate with 4,4'-isopropylidenebis[2-(2,6-dibromophenoxy) ethanol] (acetonitrile)	24-48 hrs	M_w=2,200 by GPC (details not provided)	(50)

Lipozyme IM-20 (lipase from Mucor miehei) (43.7 g/mol of diacid)	Adipic acid/1,4-butanediol (diisopropyl ether). Molecular sieves for removal of water and aqueous base extraction of polymer at midpoint.	70 hr/stage (2 stages)	M_w=4,645 by GPC analysis against PEG standards.	(56)
porcine pancreatic lipase (400 g/mol of diester)	bis(2,2,2-trichloroethyl) adipate with 1,4-butanediol in diethyl ether.	400 hrs	M_w=5,400 by GPC with 12 PS standards within 162-22,000 range. M_w = 4,100 by mass spectrometry	(42)
Lipase from M. miehei (167 g/mol of diester)	bis(2,2,2-trifluoroethyl) sebacate with 1,4-butanediol in diphenyl ether. Periodic removal of trifluoroethanol under vacuum (5 mm of Hg followed by 0.15 mm of Hg)	168 hrs	M_w = 46,400 by GPC analysis against polystyrene standards	(34)

It is important to stress that certain research groups did not seek to achieve high molecular weight polyesters in their studies so the absence of high molecular weight does not indicate a failed approach. Methods of molecular weight determination are also included for comparison.

(SOURCE: Adapted from Chaudhary et al. *Biotech. Progress*, In Press)

Table IV. Enzyme-catalyzed AA-BB type bulk polymerization between
divinyl adipate and 1,4-butanediol.

Enzyme concentration	M_w (PDI) (GPC)	M_w (PDI) (GPCV)	Reaction Extent	Acid number (mg KOH/g of polyester)	COOH equivalents per mole
0.2 wt % (0.58 g/mol of diester)	2,185 (1.6)	-	83 %	11.60	0.28
2 wt % (5.8 g/mol of diester)	17,746 (2.5)	9110 (1.45)	97.8 %	16.8	1.88
5 wt % (14.4 g/mol of diester)	23,236 (2.78)	12865 (2.38)	98.3 %	15.9	1.53
10 wt % (28.8 g/mol of diester)	Reaction vial exploded with spill of reaction substrates				

Due to poor heat removal in the incubator (air at 50 °C as the cooling medium and
low surface area for heat dissipation through glass vials), higher temperatures are
reached in the system with the extreme case of a runaway reaction at 10 wt %
enzyme. (Note that 10 wt % is < 0.01 mol % enzyme)

(SOURCE : Adapted from Chaudhary et al. *Biotech. Progress*, In press)

polyester is in the range of 50 cents/lb) *(57)*, greatly reduced enzyme content and shorter reaction times are a necessity.   Also, the total elimination of a reaction solvent and the elimination of side reactions will be necessary to produce large quantities of polyester.

Since divinyl adipate (diester monomer) has a melting point which is low enough to allow for the enzyme-catalyzed melt-phase polymerization, we have studied bulk polymerization for this system.   Reactions were performed in an incubator/shaker that was maintained at 50 °C (Table IV).   **We eliminated the solvent entirely from our system and the enzyme is dispersed within the reactants themselves.**

Table IV represents results for the most efficient (in time and enzyme concentration) biocatalytic polyester syntheses published by any group to date. There is a two order of magnitude advantage in both time and enzyme content as compared to results obtained for alternative strategies.

Since the water for hydrolysis comes from the enzyme bound water molecule, it is imperative that low water should be introduced in the reaction mixture. This can be done by reducing the enzyme concentration. However, this could lead to lower reaction rates and subsequent loss of enzyme activity.   We have circumvented the above problem by increasing the substrate concentration.

Through a series of kinetic experiments we have determined that increasing enzyme/substrate concentrations does indeed lead to rapid attainment of higher molecular weights within short times.   However the polyester (of same molecular weight) obtained at higher E/S ratio has a higher acid-content than the one obtained at low E/S content (Figure 10).

Therefore, the ratio of hydrolysis rate to transesterification rate is directly proportional to E/S ratio. Furthermore, once the polymerization mixture consists of a large fraction of acid, the condensation of acid end alcohol groups (which is also catalyzed by Novozym-435 but at a significantly diminished rate) is responsible for further growth of polymerization.   The overall effect of these processes is to reduce overall polymerization rate with increasing size and enzyme concentration.

The reaction is highly exothermic and higher condensations occur during short time periods with increasing enzyme concentration. Due to this and poor heat transfer characteristics, the initial phase of polymerization proceeds in an adiabatic manner.   Obviously more significant temperature surges are observed with higher enzyme concentrations, and are responsible for faster polymerization despite higher acid formation.

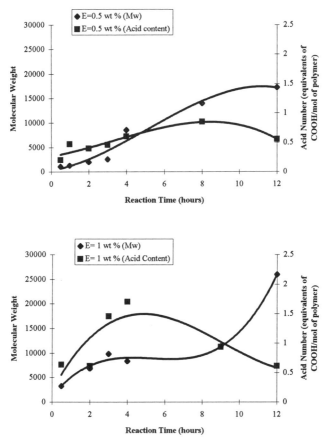

Figure 10. Effect of E/S ratio on polymer molecular weight and acid content for bulk polymerization between DVA and 1,4-BD.
SOURCE : Adapted from Chaudhary et al. *Biotech. Progress*, In press.

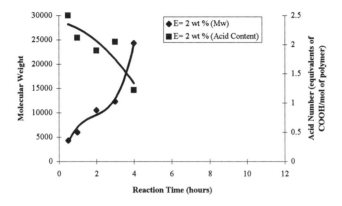

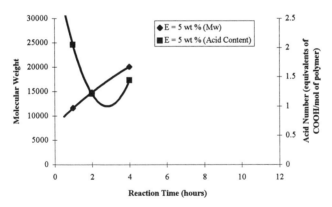

Figure 10. *Continued.*

Solvent Engineering of Polyester Properties

Dilution of the bulk reaction mixture with a polar solvent such as THF was performed to investigate the role of organic solvent on controlling molecular weight and acidity of the polymer (Figure 11). It is interesting to note that whereas there is no significant effect of solvent dilution on polymer acid content (it is important to use acid content as the equivalents of carboxylic acid groups per mole of polymer molecule since the acid number in mequivalents of KOH/g of polymer can be misleading when comparing polymers of different molecular weights), molecular weights of the polymer obtained at lower concentrations are lower. Therefore, increasing the amount of solvent to the system results in the onset of hydrolysis at lower molecular weights. This is a combined result of the facilitated release of enzyme-bound water in the presence of hydrophilic solvent such as THF alongwith the reduced adiabacity of the reaction because of heat quenching by the solvent.

Solvent fraction can therefore be manipulated to exercise molecular weight control (maintaining identical acid content) for DVA/BD system. Previously, we were also able to regulate molecular weight through precipitation in a pressure tunable supercritical environment for a bis(2,2,2-trichloroethyl) adipate/1,4-BD system.

Effect of Temperature

Because of the sensitive nature of the biocatalyst, the temperature range for operation of biocatalytic process is relatively narrow. Increasing the temperature will increase enzyme activity but decrease enzyme stability. Morrow and coworkers suggested that increasing temperature can also alter relative rates of hydrolysis and transesterification and at the same time increase the ease of removal of alcohol byproduct. Therefore it was not possible to identify if the higher DP at higher temperature was related to increased transesterification or shifted equilibrium.

We performed reactions at different fractions of THF. Both sets of reactions underwent similar initial temperature jumps but after the initial phase, in one case the temperature was held constant at 25 °C and at 50 °C in other case by fixing the incubator temperature. Similar molecular weight averages were obtained with lower substrate concentrations leading to lower molecular weights. Therefore, decreasing the reaction temperature after the adiabatic phase does not affect either molecular weight or acid number. Therefore the temperature effect during adiabatic phase seems to be more important factor for our system. We are currently trying to obtain useful kinetic information from the time-temperature profile.

Enzyme Stability/Recyclability:

Very few attempts have been made to determine the stability and recyclability of enzyme for AA-BB type polymerization. We have reported earlier that adding fresh aliquots of porcine pancreatic lipase periodically to organic solvent did not increase polymer molecular weight indicating that loss of enzyme activity at long

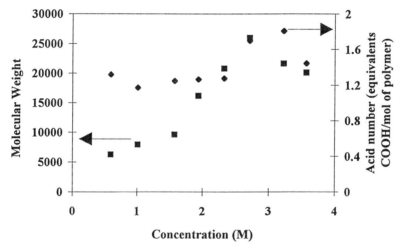

Figure 11. Effect of THF concentration on polyester molecular weight and acid content for bulk polymerization between DVA and 1,4-BD. ([E] = 5 wt %, Time = 4 hrs).

SOURCE : Adapted from Chaudhary et al. *Biotech. Progress*, In press.

reaction times in organic environment was not responsible for the termination of polymerization. Since AA-BB type reactions have been performed by different researchers using lipases from different sources for varying periods of time, it is not possible to comment on the stability of an individual enzyme in one particular study. Performing reactions for long times involves a danger of loss of enzyme activity, which can limit the polymerization growth.

For our system though significantly low time periods were associated with rapid temperature jumps. The effect of later will be to reduce stability. We recovered and recycled the enzyme obtained from subsequent experiments and measured the synthetic utility of the spent catalyst after every batch. We have observed good recyclability of Novozym-435 for further catalysis (Figure 12).

Kinetic Features of Enzyme Catalysis

The kinetics of polymerization are of prime interest from two viewpoints. The practical synthesis of polymers requires a knowledge of the kinetics of the polymerization reaction. From the theoretical viewpoint, significant differences between step and chain polymerizations reside in large part in their kinetic features *(23)*.

Flory's Kinetic analysis. For conventional chemically catalyzed step polymerization (e. g. between diacid and diol), polymerization proceeds by a gradual increase in polymer molecular weight. Monomers combine together first to form dimer, dimer forms trimer by reaction with either diacid or diol monomer and tetramer by reacting with itself. Polymerization proceeds in a stepwise manner with the disappearance of monomer very early in the reaction. Kinetic analysis of the polymerization *(58)* is greatly simplified by assuming that (i) the reactivities of both functional groups of a bifunctional group (both hydroxyls of a diol) are the same (ii) the reactivity of one functional group of a bifunctional reactant is the same irrespective of whether the other functional group has reacted and (iii) Reactivity of a functional group is independent of the size of the size of the molecule to which it is attached. These simplifying assumptions make the kinetics of step polymerization identical to those for the analogous small molecule reaction (for example esterification of acetic acid with ethanol).

Flory (1953) studied reactions of certain small nonpolymeric molecules to verify the validity of these assumptions. Both theoretical *(59)* and experimental justifications of these assumptions have been offered (Exceptions to this occur when the reactivities of the functional groups are very high and/or the molecular weights are very high, polymerization becomes diffusionally controlled in these cases).

For esterification between homologous series of carboxylic acids *(60)*, decreased reactivity with increased molecular size (at very small size) was observed with limiting value reached at small size after which it remains constant. Similar

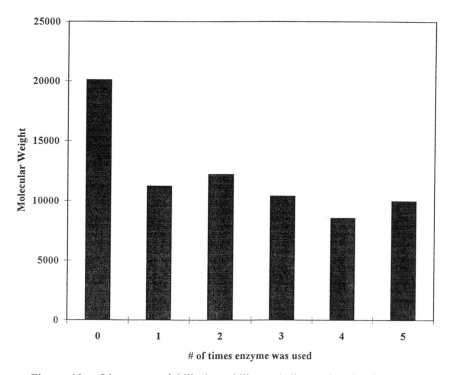

Figure 12.   Lipase recyclability/reusability.   (All reaction batches were performed for 4 hours of reaction.   Initial amount of enzyme corresponds to 5 wt % enzyme.   After the reaction, enzyme was washed with THF to prevent loss of polymer bound to enzyme).
SOURCE : Adapted from Chaudhary et al. *Biotech. Progress*, In press.

results were also obtained for polyesterification of sebacoyl chloride and α, ꭃ-alkanediols *(61)*. It is important to note that above experiments were performed on compounds within a homologous series (that differing from each other by methylene groups within the backbone).

Kinetics of enzyme-catalyzed polytransesterification. Kinetics of transesterification *(62)* and esterification *(63)* by hydrolases have been studied previously in organic solvents *(64)* and supercritical fluids *(65)*. These studies have shown that the transesterification follows acyl-enzyme *(66,67)* or ping pong bi bi mechanism *(68)* with water (for the corresponding hydrolysis kinetics) replaced by the nucleophile in the deacylation step. Microscopic rate constants for an enzyme catalyzed transesterification using an added nucleophilic method (69) and kinetic parameters have been determined previously from initial rate studies.

Since preferred disappearance of one of the starting monomers (diester or diol) has been observed by different research groups during enzymatic polymerization *(32,38,* Chaudhary et. al. *Biotech. Bioengg.* Submitted), the assumption of equal reactivity does not hold for an enzyme catalyzed synthesis.

Based on this information, we have simulated the growth of enzyme-catalyzed polymerization using a computational model (Chaudhary et. al., *Annal. N. Y. Acad. Sci.,* In press). This model consists of a complex network of consecutive and parallel reactions. The change in concentration of all individual species can be calculated from the kinetic model.

$$\frac{d(species)}{dt} = \left(\sum Formation \right)_{species} - \left(\sum Disappearence \right)_{species}$$

Insight into the how the features of enzyme catalysis will manifest into polymer molecular weight distribution can be obtained by performing mathematical simulations and our results will be published in coming years.

Directions for the Future.

To summarize therefore, enzymatic polymerization in organic solvents is an interesting phenomena. Activation of diester, length of diol chain, enzyme/substrate ratio, enzyme stability, activity and the solvent choice are different parameters that influence the polyester molecular weight and the functionality.

Predictive kinetic model will help to identify the controlling steps in the enzymatic polymerization. This information will be valuable in designing the synthesis parameters *apriori* for controlling the polymerization reaction.

Acknowledgments. This work has been supported by research grants from National Science Foundation, Bayer Corporation and Army Research Office. We also like to acknowledge Micromass Inc. for providing access to use MALDI Mass spectrometry.

Literature Cited.

(1) *Polyesters in Encyclopedia of Polymer Science and Engineering*, **1992**, 3rd edition, Vol. 12, 1-50.

(2) *Polyesters in Ulmann's Encyclopedia of Industrial Chemistry*, Vol. A21, 227.

(3) Johns, D. B., Lenz, R. W., Luecke, A., , *Lactones* In *Ring-Opening polymerization;* Ivin, K. J., Saegusa, T., Elsevier, London, **1984**, Vol. 1.

(4) Lundberg, R. D., Cox, E. F., *Lactones* in *Ring-Opening polymerization*, Frisch, K. C., Reegen, S. L., Reegen eds. Marcel Dekker, New York, **1969**.

(5) Sanders, J. H., Frisch, K. C. *Reactions of isocyanates and isocyanate derivatives in Polyurethanes : Chemistry and Technology. Part I Chemistry* John Wiley & Sons, Inc., New York, **1992**, pp 63-128.

(6) Tirrell, J. G., Fournier, M. J., Mason, T. L., Tirrell, D. A., *C & E News*, **1994**, Dec 19, pp 40-51.

(7) McDaniel, R., Ebert-Khosla, S., Hopwood, D. A., Khosla, C., *Science*, **1993**, *262*, pp 1546-1550.

(8) Yoshikawa, E., Fournier, M. J., Mason, T. L., Tirrell, D. A., *Macromolecules*, **1994**, *27*, pp 5471-5475.

(9) Kothakota, S., Mason, T. L., Tirrell, D. A., Fournier, M. L., *J. Am. Chem. Soc.*, **1995**, *117*, pp 536-537.

(10) Krejchi, M. T., Atkins, E. D. T., Waddon, A. J., Fournier, M. L., Mason T. L., Tirrell, D. A., *Science*, **1994**, *265*, 1427-1432.

(11) Tokiwa, Y., Ando, T., Suzuki, T., Takeda, T., *Polymeric Materials : Science and Engineering*, **1990**, *Vol. 70*.

(12) Mobley, D. P., *Plastics from Microbes: Microbial synthesis of polymers and polymer precursors*, **1994**, Munich: Hanser Publications.

(13) Kim, Y. B., Lenz, R. W., Fuller, R. C., *Macromolecules*, **1992**, *25*, pp 1852-1857.

(14) Anderson, J. A., Dawes, E. A., *Microbiol. Rev.* **1990**, *54*, 450.

(15) *Biocatalysts for Industry*, Dordick, J. S., Plenum press, New York, **1991**.

(16) *Enzymatic Reaction Mechanisms*, Walsh, C.,W. H. Freeman & Co., San Francisco, **1979**.

(17) *Biotransformations in Organic Chemistry- A Textbook*, Faber, K., 2nd edition, Springer-Verlag, Berlin, **1995**.

(18) *Enzyme Nomenclature*, International Union of Biochemistry and Molecular Biology, Academic Press, New York, **1992**.

(19) Kindel, S., Technology, **1981**, *1*:62.

(20) Dordick, J. S., Marletta, M. A., and Klibanov, A. M., *Biotechnol. Bioeng.* **1987**, *30*, 31-36.

(21) *Enzyme catalyzed polyester synthesis and fractionation using supercritical fluids*, Chaudhary, A. K., **1994**, M. S. thesis, University of Pittsburgh, Pittsburgh.

(22) Dordick, J. S., *TIBTECH*, **1992**, *10*, 287-293.

(23) Odian, G., *Principles of Polymerization*, 2nd ed., John Wiley and Sons, New York, **1981**.

(24) Gutman, A. L., Oren, D., Boltanski, A., Bravdo, T., *Tetra. Lett.*, **1987**, *28*, 44, pp 5367-5368.

(25) Knani, D., Gutman, A. L., Kohn, D. H., *J. Poly. Sci. Part A: Poly. Chem.*, **1993**, *31*, pp 1221-1232.

(26) Knani, D., Kohn, D. H., *J. Poly. Sci., Part A: Polym. Chem.*, **1993**, *31*, 2887-2897.

(27) Uyama, H., Kobayashi, S., *Chem. Lett.*, **1993**, 1149.

(28) MacDonald, R. T., Pulapura, S. K., Svirkin, Y. Y., Gross, R. A., Kaplan, D., Akkara, J., Swift, G., Wolk, S., *Macromolecules*, **1995**, 28, 73-78.

(29) Okumara, S., Iwai, M., Tominaga, Y., Agric. Biol. Chem. **1984**, *48* (11), pp 2805-2808.

(30) Margolin, A. L., Crenne, J-Y., Klibanov, A. M., *Tetra. Lett.*, **1987**, *28*, 15, pp 1607-1610.

(31) Wallace, J. S., C. J., *J. Poly. Sci. Part A : Poly. Chem.*, **1989**, *27*, pp 3271-3284.

(32) Morrow, C. J., Wallace, J. S., Bybee, G. M., Reda, K. B., Williams, M. E., *Materials Research Society Symposium Proceedings*, **1990**, *174*.

(33) Wallace, J. S., Morrow, C. J., *J. Poly. Sci. Part A : Poly. Chem.*, **1989**, *27*, 2553-2567.

(34) Linko, Y-Y., Wang, Z-L., Seppala, J., *Enz. Microb. Technol.*, **1995**, *17*, pp 506-511.

(35) Wang, Y. F., Lalonde, J. J., Momongan, M., Bergbreitter, D. E., Wong, C. H., *J. Am. Chem. Soc.*, **1988**, *110*, 7200.

(36) Ghogare, A. D., Kumar, S., *J. Chem. Soc. Chem. Commun.*, **1989**, 1533.

(37) Uyama, H., Kobayashi, S., *Chem. Lett.*, **1994**, pp 1687-1690.

(38) Athavale, V. D., Gaonkar, S. R., *Biotech. Lett.*, **1994**, *16* (2) pp 149-154.

(39) Klempier, N., Faber, K., Griengl H, *Synthesis*, **1989**, 933.

(40) Linko, Y-Y., Seppala, J., *CHEMTECH*, 1996, August, pp 25-31.

(41) *Product Information sheet, Novo Nordisk Bioindustrials Ltd.* Christensen, A. H., Enzymatic resolution of aminoalcohols, Master's thesis, The technical University of Denmark in Refrence summary.

(42) Chaudhary, A. K., Beckman, E. J., Russell, A. J., *J. Am.Chem. Soc.,* 1995, *117*, pp- 3728-3733.

(43) Bahr, U., Deppe, A., Karas, M., Hillenkamp, F., *Anal. Chem.*, **1992**, *64*, pp 2866-2869.

(44) Blais, J. C., Tessier, M., Bolbach, G., Remaud, B., Rozes, L., Guittard, J., Brunot, A., Marechal, E., Tabet, J. C., *Int. J. Mass Spectro. and Ion processes*, **1995**, *144*, pp 131-138.

(45) Chait, B. T., Wang, R., Beavis, R. C., Kent, S. B. H., *Science*, **1993**, *262*, pp
(46) Danis, P. O., Karr, D. E., Simonsick, W. J., Wu, D. T., *Macromolecules*, **1995**, *28*, 1229.
(47) Abate, R., Ballisteri, A., Montaudo, G., Garozzo, D., Impaliomeni, G., Critchley, G., Tanaka, K., *Rapid Commun. Mass Spectrom.*, **1995**, *7*, 1033.
(48) Montaudo, G., Montaudo, M. S., Puglisi, C., Samperi, F., *Rapid Commun. Mass Spectrom.*, **1994**, *8*, 981.
(49) Montaudo, G., Montaudo, M. S., Puglisi, C., Samperi, F., *Rapid Commun. Mass Spectrom.*, **1995**, *9*, 453.
(50) Geresh, S., Gilboa, Y., *Biotech. Bioengg.*, **1990**, 36, pp 270-274.
(51) Geresh, S., Gilboa, Y., *Biotech. Bioengg.*, **1991**, 37, pp 883-888.
(52) Geresh, S., Gilboa, Y., Abrahami, S., Bershadsky, A., *Poly. Engg and Science*, **1993**, *33* (5), 311-315.
(53) Brazwell, E. M., Filos, D. Y., Morrow, C. J., *J. Poly. Sci., Part A: Poly. Chem.*, **1995**, *33*, pp 89-95.
(54) Chaudhary, A. K., Critchley, G., Diaf, A., Beckman, E. J., Russell, A. J., *Macromolecules*, **1996**, *29(6)*, pp 2213-2221.
(55) Morrow, C. J., *MRS Bulletin*, **1992**, November.
(56) Binns, F., Roberts, S. M., Taylor, A., Williams, C. F., *J. Chem.Soc. Perkin Trans.*, **1993**, 1, 899-904.
(57) Palmisano, A., Pettigrew, C. A., *Bioscience*, **1992**, *42*(9): 680-685.
(58) Flory, P. J., *Principles of Polymer Chemistry*, Cornell University Press, Ithaca, New York, **1953**.
(59) Rabinowitch, E., *Trans. Faraday Soc.*, **1937**, *33*, 1225.
(60) Bhide, B. V., Sudborough, J. J., *J. Indian Inst. Sci.*, 8A, 89, 1925
(61) Ueberreiter, K., Engel, M., *Makromol. Chem.*, **1977**, 178, 2257.
(62) Zaks, A., Klibanov, A. M., *J. Biol. Chem.*, *263*, 3194-3201.
(63) Marty, A., Chulalaksananukul, W., Willermot, R. M., Condoret, J. S, *Biotech. Bioengg.*, **1992**, *39*, 273-280.
(64) Chulalaksananukul, W., Condoret, J. S., Delorme, P., Willemot, R. M., *FEBS Lett.*, **1992** 276 (1,2):181-184.
(65) *Biocatalysis and Bioseparations in Supercritical Fluids*, Kamat. S. V., **1996**, Ph. D. Dissertation, University of Pittsburgh, Pittsburgh.
(66) Adams, K.A.H., Chung, S. H., Klibanov, A. M., *J. Am. Chem. Soc.*, **1991**, 112:25.
(67) Chatterjee, S., Russell, A. J., *Biotech. Progress*, **1993**, *8*, pp 256-258.
(68) *Biocatalysis and Bioseparations in Supercritical Fluids*, Kamat, S. V., Ph. D. Dissertation, **1996**, University of Pittsburgh, Pittsburgh.
(69) Chatterjee, S., Russell, A. J., *Biotech Bioeng.*, **1992**, Vol. 40, 1069-1077.

Chapter 3

Enzymatic Polymerization for Synthesis of Polyesters and Polyaromatics

Shiro Kobayashi and Hiroshi Uyama

Department of Materials Chemistry, Graduate School of Engineering, Kyoto University, Kyoto 606–01, Japan

Enzymatic syntheses of polyesters and polyaromatics were investigated. Ring-opening polymerization of lactones with different ring-size proceeded through lipase catalysis. One-shot synthesis of methacryl-type macromonomer was achieved by the polymerization in the presence of vinyl methacrylate. Lipase also catalyzed polycondensation of divinyl adipate with glycols and poly(addition-condensation) of succinic anhydride with glycols to produce aliphatic polyesters under mild reaction conditions. Peroxidase-catalyzed polymerization of phenol derivatives afforded a new class of polyphenols showing high thermal stability. Enzymatic synthesis of poly(1,4-oxyphenylene) was performed by oxidative polymerization of 3,5-dimethy-4-hydroxybenzoic acid monomer catalyzed by peroxidase or laccase involving elimination of carbon dioxide and hydrogen from the monomer. Peroxidase catalyzed regioselective polymerization of p-alkoxyphenols, leading to the formation of poly(phenylene oxide). Polyphenol particles were obtained by peroxidase-catalyzed dispersion polymerization in the presence of poly(vinyl methyl ether) stabilizer.

Enzyme-catalyzed reactions in organic solvents have been increasingly important in organic synthesis (1-3). Polymerizations catalyzed by enzymes (enzymatic polymerizations) received little attention until several years ago because such specific properties of enzymes had not been fully utilized in most of these polymerizations (4,5). Recently, however, the synthesis of cellulose via a non-biosynthetic path has successfully been achieved by the enzymatic polymerization of β-cellobiosyl fluoride in an aqueous organic solvent, using cellulase as catalyst (6). This methodology has been applied to synthesis of natural and non-natural polysaccharides (7,8). We have systematically explored enzymatic syntheses of polyesters and polyaromatics for the last decade. This chapter deals with recent development on enzymatic polymerizations to these polymers.

Enzymatic Synthesis of Polyesters

Enzymatic Ring-Opening Polymerization of Lactones. Most of the enzymatic polymerizations hitherto reported were of polycondensation type. Recently, we have expanded enzymatic polymerizations to lipase-catalyzed ring-opening polymerization and copolymerization of lactones. Small- (4-membered), medium-size (6- and 7-membered) lactones as well as macrolides (12-, 13-, and 16-membered) were polymerized through lipase catalysis to produce the corresponding polyesters (Figure 1). Polymerization results are summarized in Table I.

Table I. Enzymatic Polymerization of Lactones Catalyzed by Lipase[a]

Monomer	Lipase[b]	Temp. (°C)	Time (h)	Conv.[c] (%)	Mn[c] (x10⁻³)	Mw/Mn[c]
β-PL	Lipase PF	60	120	84 [d]	0.66	1.6
δ-VL	Lipase PF	60	120	87	1.7	3.8
ε-CL	Lipase CC	60	240	92	1.9	2.0
ε-CL	Lipase PF	60	240	71	7.0	2.2
ε-CL	Lipase PF	75	240	94	7.7	2.4
ε-CL	Lipase PF	75	480	99	11.9	2.3
UDL	Lipase CC	60	240	95	11.7	2.4
UDL	Lipase CC	75	240	95	25.2	2.2
UDL	Lipase PF	45	120	96	2.9	2.5
UDL	Lipase PF	60	48	98	8.5	2.4
UDL	Lipase PF	75	48	98	19.5	2.5
UDL	Lipase PF	75	240	100	22.8	2.6
DDL	Lipase CC	75	120	99	13.0	2.8
DDL	Lipase PF	75	240	100	11.4	4.3
PDL	Lipase CC	75	240	65	16.2	2.4
PDL	Lipase PF	75	120	96	7.2	2.7

[a] Polymerization was carried out in bulk. [b] Lipase PF: *Pseudomonas fluorescens* lipase; Lipase CC: *Candida cylindracea* lipase. [c] Determined by GPC. [d] Isolated yield.

The polymerization was carried out in bulk. From β-propiolactone (β-PL), an oligomer with molecular weight of less than 1×10^3 was obtained by using *Pseudomonas fluorescens* lipase (lipase PF) as catalyst (9). Lipase PF also catalyzed the ring-opening polymerization of δ-valerolactone (δ-VL) and ε-caprolactone (ε-CL), yielding the polymer having the molecular weight of several thousands (10). *Candida cylindracea* lipase (lipase CC) was available as catalyst for the polymerization of ε-CL. In the polymerization of the medium-size lactones without the enzyme (control experiment), all the monomers were recovered unreactedly, indicating that the polymerization proceeds via enzyme catalysis.

Macrocyclic esters (macrolides) have virtually no ring strain, and hence, show similar reactivities, *e.g.*, in alkaline hydrolysis with acyclic fatty acid alkyl esters. Anionic polymerizability of macrolides was much lower than that of ε-CL possessing high strain in ring. We have examined the polymerization of 12-dodecanolide (DDL) catalyzed by various lipases of different origin at 60 °C for 120 h (Table II) (11). Among lipases examined, lipases derived from *Pseudomonas* family (lipases PA, PC, and PF), lipase CC, and porcine pancreas lipase (PPL) exhibited high catalytic activities for the polymerization; the monomer was quantitatively consumed by using these enzymes. In other lipases, less or no catalytic activities for the polymerization was observed. Other hydrolases having esterase activities such as protease and amino acylase did not induce the polymerization of DDL. These results indicate that the polymerization behavior greatly depended on the type and origin of enzyme.

Table II. Enzyme Screen for Enzymatic Polymerization of DDL[a]

Lipase			Polymer	
Origin	Code	Conv.[b] (%)	Mn^b ($\times 10^{-3}$)	Mw/Mn^b
Aspergillus niger	Lipase A	12	1.2	1.2
Candida cylindracea	Lipase CC	99	7.3	2.3
Penicillium roqueforti	Lipase PR	0	---	---
Pseudomonas aeruginosa	Lipase PA	99	7.7	3.4
Pseudomonas cepacia	Lipase PC	100	5.6	2.3
Pseudomonas fluorescens	Lipase PF	99	4.0	2.5
Rhizopus delemer	Lipase RD	<5	---	---
porcine pancreas	PPL	99	3.2	2.0
control	---	0	---	---

[a] Polymerization was carried out in bulk. [b] Determined by GPC.

Other macrolides, 11-undecanolide (UDL) and 15-pentadecanolide (PDL), were subjected to the lipase-catalyzed polymerization (Table I) (*12,13*). The polymerization of the macrolides at 75 °C produced the polymer with the molecular weight of more than 1×10^4. The highest molecular weight (2.5×10^4) was achieved by the polymerization of UDL at 75 °C by lipase CC. From ^1H and ^{13}C NMR analyses, the terminal structure of the polylactone was found to be of carboxylic acid at one end and of alcohol at the other terminal except that of poly(β-PL). Lipase catalysis also induced the enzymatic copolymerization of lactones. From the combination of δ-VL and ϵ-CL, the random copolymer was obtained (*14*).

Table III summarizes dipole moment and reactivities of lactones with different ring size. The dipole moment is shown as an indication of their ring strain. The values of macrolides are lower than that of ϵ-CL and close to that of an acyclic ester (butyl caproate). The rate constants of these macrolides in alkaline hydrolysis and propagation of anionic ring-opening polymerization are much smaller than those of ϵ-CL. These data imply that these macrolides possess much lower ring strain, and hence, exhibit less anionic reactivity and polymerizability than ϵ-CL.

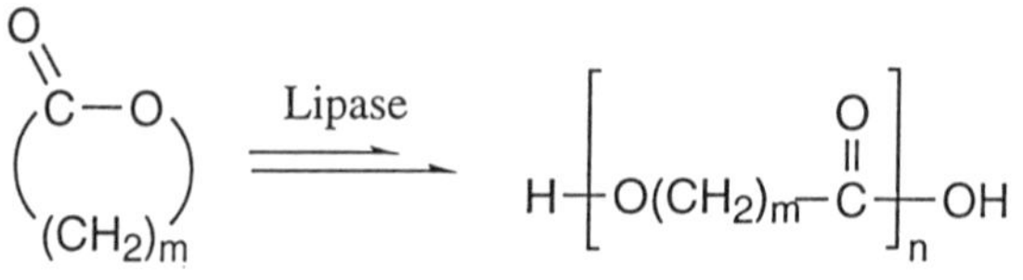

$$\left(\begin{array}{c} O \\ \backslash\backslash \\ C-O \\ (CH_2)_m \end{array}\right) \quad \xrightarrow{\text{Lipase}} \quad H \left[O(CH_2)_m \overset{O}{\underset{\parallel}{C}} \right]_n OH$$

m=2: β-PL   m=4: δ-VL   m=5: ϵ-CL

m=10: UDL   m=11: DDL   m=14: PDL

Figure 1. Lipase-catalyzed ring-opening polymerization of lactones (β-PL: β-propiolactone, δ-VL: δ-valerolactone, ϵ-CL: ϵ-caprolactone, UDL: 11-undecanolide. DDL: 12-dodecanolide, PDL: 15-pentadecanolide).

Table III. Dipole Moments and Reactivities of Lactones

Lactone	Dipole Moment (μ)	Alkaline Hydrolysis[a] ($M^{-1} \cdot s^{-1}$, $\times 10^4$)	Propagation[b] (s^{-1}, $\times 10^3$)	R_{max}[c] ($mol \cdot L^{-1} \cdot h^{-1}$)
		Rate Constant		
ε-CL	4.45	2550	120	0.040
UDL	1.86	3.3	2.2	0.31
DDL	1.86	6.0	15	0.17
PDL	1.86	6.5	---	0.25
Butyl Caproate	0.75	8.4	---	---

[a] Alkaline: NaOH. Measured in 1,4-dioxane/water (60/40 vol%) at 0 °C. [b] Measured using NaOMe initiator (6 mol%) in THF at 0 °C. [c] Measured using lipase PF catalyst in bulk at 60 °C.

Figure 2 shows time-conversion curves in the polymerization of ε-CL, UDL, and DDL with lipase PF catalyst at 60 °C. The polymerization of the macrolides proceeded much faster than that of ε-CL. In order to evaluate the lipase-catalyzed polymerization rate quantitatively, an apparent maximum rate (R_{max}) was determined from the time-conversion curve (Table III). The R_{max} values of the macrolides were larger than that of ε-CL. This tendency is opposite to that of anionic polymerization, suggesting that the lipase catalyst shows much higher catalytic activity toward the macrolides than it does for ε-CL.

We proposed that the lipase-catalyzed polymerization of lactones proceeds as follows (Figure 3). Lipase-catalyzed reactions proceed via an acyl-enzyme intermediate and its catalytic site is a serine-residue. In the present polymerization, the key step is the reaction of the lactone with lipase involving the ring-opening of the lactone to produce the acyl-intermediate (enzyme-activated monomer, EM). The initiation is a nucleophilic attack of water, which is perhaps contained in the enzyme, on the acyl carbon of the intermediate, yielding ω-hydroxylcarboxylic acid (n=1), which is regarded as the shortest propagating species. The intermediate is nucleophilically attacked by the terminal hydroxyl group of the polymer in the propagation stage, leading to the formation of one-unit-more elongated polymer chain. The kinetic study showed that the rate-determining step of the over-all polymerization is the formation of the enzyme-activated monomer, therefore, the present polymerization proceeds via a "monomer-activated mechanism".

Then, we have made Michaelis-Menten kinetics of the polymerization (*15*). For kinetics of a two-substrate enzymatic reaction, there is a question which substrate is bound first. In the lipase-catalyzed polymerization of lactones in the presence of alcohol, the concentration of the hydroxyl group is considered to be constant in the early stage of the polymerization. Therefore, we paid attention to the stage of the acyl-enzyme intermediate formation and determined the kinetic parameters $K_{m(lactone)}$ and $V_{max(lactone)}$ in the presence of an appropriate amount of 1-octanol.

These parameters were determined by Hanes-Woolf plots. The present polymerization followed Michaelis-Menten kinetics: linearity was observed for both monomers in the Hanes-Woolf plot. The parameters obtained are summarized in Table IV. Reciprocal of $K_{m(lactone)}$ of DDL was about one-half as that of ε-CL, on the other hand, $V_{max(lactone)}$ of DDL was more than three times larger than that of ε-CL, indicating that the larger polymerizability of DDL through lipase catalysis is mainly due to the larger reaction rate (R_{max}), but not to the binding abilities. These results suggest

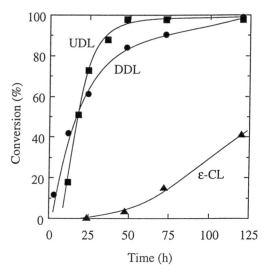

Figure 2. Time-conversion curves in the enzymatic polymerization of lactones with different ring size using lipase PF catalyst at 60 °C in bulk.

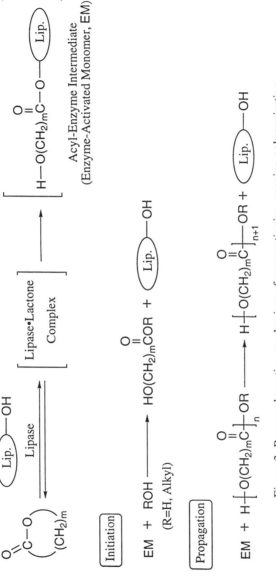

Figure 3. Proposed reaction mechanism of enzymatic ring-opening polymerization of lactones.

that the reaction process of the lipase-lactone complex to the acyl-enzyme intermediate is the key step in the present polymerization.

Table IV. Michaelis-Menten Kinetic Parameters in Ring-Opening Polymerization of Lactones Catalyzed by Lipase PF[a]

Lactone	$K_{m(lactone)}$ $(mol \cdot L^{-1})$	$V_{max(lactone)}$ $(x10^2, mol \cdot L^{-1} \cdot h^{-1})$
ε-CL	0.61	0.66
DDL	1.1	2.3

[a] Polymerization was carried out using lipase PF (200 mg) catalyst in the presence of 1-octanol (0.03 M) in i-propyl ether (10 mL) at 60 °C.

One-Shot Synthesis of Polyester Macromonomer and Telechelics. One-shot acylation of the polymer terminal was examined by the polymerization of DDL using lipase PF catalyst in the presence of vinyl esters, which are often used as acylating agents through lipase catalysis in organic syntheses. Enzymatic synthesis of methacryl-type polyester macromonomer was attempted by the polymerization of DDL in the presence of vinyl methacrylate (Figure 4) (16). By using 12.5 or 15 % vinyl methacrylate based on DDL, the polymerizable group was quantitatively introduced at the terminal.

The macromonomer formation is probably explained as follows. During the polymerization of DDL, the acyl-enzyme intermediate is formed from the vinyl ester and lipase, which is subjected to the reaction with the terminal hydroxyl group of the propagating polymer, leading to the macromonomer. Vinyl 10-undecenoate afforded the ω-alkyenyl-type macromonomer.

These macromonomers were obtained by a convenient, one-shot procedure. This process was a new type of macromonomer syntheses, since the vinyl ester acting as terminator was present from the beginning of the reaction. Furthermore, this system can be applied to synthesis of a telechelic polyester having a carboxylic acid at both ends by the addition of divinyl sebacate (Figure 5).

Enzymatic Polycondensation of Bis(enol ester)s with Glycols. For enzymatic synthesis of polyesters by polycondensation, activated diesters, typically 2,2,2-trifluoroethyl and 2,2,2-trichloroethyl esters, are often used. However, this process involves the transesterification of the product polymer with the leaving alcohol, resulting in the oligomer formation. We have paid attention to the high reactivity of enol esters in the lipase-catalyzed reactions and used a bis(enol ester) as a monomer for the polyester synthesis.

Enzymatic polymerization of vinyl adipate with 1,4-butanediol proceeded in i-propyl ether at 45 °C in the presence of lipase PF (17). After 48 h, the aliphatic polyester was obtained in 50 % isolated yield and its molecular weight was 6700. Divinyl sebacate was also available as monomer. Chain length of 1,ω-alkylene glycol affected the yield and molecular weight of the polymer. Under the same reaction conditions, no polymer formation was observed from adipic acid or diethyl adipate, indicating that the bis(enol ester) was very reactive toward the lipase catalyst to produce aliphatic polyesters.

Enzymatic Copolymerization of Lactones, Divinyl Esters, and Glycols. As described above, lipase-catalyzed polycondensation and ring-opening

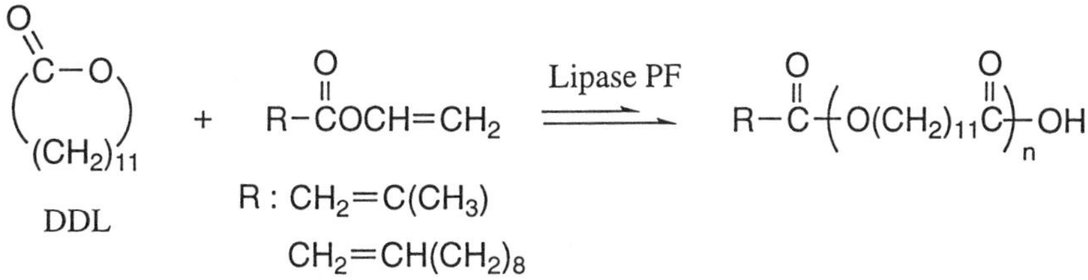

$$
\begin{pmatrix} \overset{\displaystyle O}{\overset{\|}{C}} - O \\ (CH_2)_{11} \end{pmatrix} + R - \overset{O}{\overset{\|}{C}}OCH=CH_2 \xrightarrow{\text{Lipase PF}} R - \overset{O}{\overset{\|}{C}} \left(O(CH_2)_{11}\overset{O}{\overset{\|}{C}} \right)_n OH
$$

DDL

R : $CH_2=C(CH_3)$

$CH_2=CH(CH_2)_8$

Figure 4. One-shot synthesis of polyester macromonomers by enzymatic polymerization of DDL in the presence of vinyl esters.

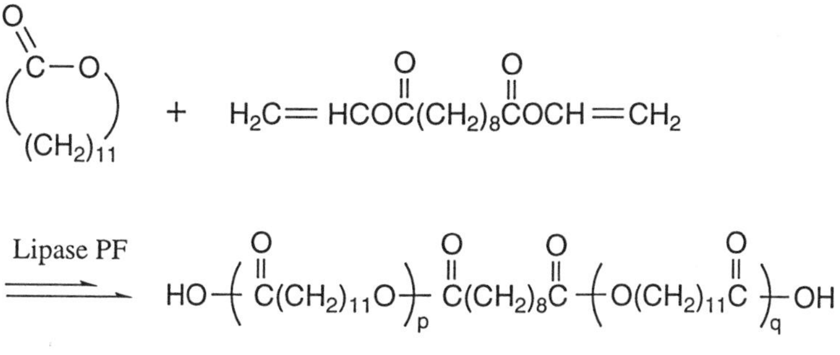

$$
\begin{pmatrix} \overset{\displaystyle O}{\overset{\|}{C}} - O \\ (CH_2)_{11} \end{pmatrix} + H_2C = HCO\overset{O}{\overset{\|}{C}}(CH_2)_8\overset{O}{\overset{\|}{C}}OCH=CH_2
$$

$$
\xrightarrow{\text{Lipase PF}} HO \left(\overset{O}{\overset{\|}{C}}(CH_2)_{11}O \right)_p \overset{O}{\overset{\|}{C}}(CH_2)_8\overset{O}{\overset{\|}{C}} \left(O(CH_2)_{11}\overset{O}{\overset{\|}{C}} \right)_q OH
$$

Figure 5. One-shot synthesis of telechelic polyester having a carboxylic acid at both ends by enzymatic polymerization of DDL in the presence of divinyl sebacate.

polymerization are believed to proceed via the same reaction intermediate (an acyl-enzyme intermediate). This encouraged us to combine both polymerization reactions through enzyme catalysis to produce polyester copolymers. *Pseudomonas cepacia* lipase (lipase PC) induced the copolymerization of macrolides, divinyl esters, and glycols (*18*). The copolymerization of PDL, divinyl sebacate, and 1,4-butanediol in *i*-propyl ether at 60 °C for 72 h produced the copolymer in 80 % yield, whose molecular weight was 6.5×10^3.

From the micro-structural analysis of the copolymer by ^{13}C NMR, the present resulting product was found to be not a mixture of the homopolymers, but a copolymer from the monomers. These data indicate that different modes of polymerization, ring-opening polymerization and polycondensation simultaneously occurred in one-pot, leading to the formation of the copolyester in the present enzymatic copolymerization.

Enzymatic Poly(addition-condensation) of Cyclic Acid Anhydride and Glycols. A new type of enzymatic polymerization, poly(addition-condensation) of cyclic acid anhydride and glycols involving ring-opening of the monomer, has been developed (Figure 6). The polymerization of succinic anhydride with 1,8-octanediol using lipase PF catalyst proceeded at room temperature to produce the corresponding polymer with molecular weight of several thousands (*19*).

Preparation of Immobilized Lipase Showing High Catalytic Activity toward Enzymatic Polymerization. In enzyme-catalyzed reactions and polymerizations in organic solvents, a powdery enzyme is often suspended directly in such media. Therefore, a large amount of the enzyme is necessary because of the heterogeneous reaction. We have dramatically reduced the enzyme amount necessary for the enzymatic polymerization of macrolides by using an immobilized lipase (*20*). The immobilization was carried out by lyophilization of the buffer solution (pH 8) of lipase PF in the presence of a Celite and lyoprotectants sugars such as sucrose, glucose, and poly(ethylene glycol). In using this immobilized enzyme, 1 weight % of the net lipase PF based on the monomer was sufficient for the polymerization, whereas we have often used 20-50 % of the bulk enzyme as catalyst to produce the polymeric materials. These data indicate that the immobilized lipase adsorbed on a Celite showed the high catalytic activity for the polymerization of macrolides. This immobilized enzyme may be applicable as catalyst for enzymatic synthesis of polyesters.

Enzymatic Synthesis of Polyaromatics

Peroxidase-Catalyzed Oxidative Polymerization of Phenol Derivatives. Peroxidase is an enzyme which catalyzes the oxidation of a donor to an oxidized donor by the action of hydrogen peroxide, liberating two water molecules. A peroxidase-catalyzed oxidative reaction of various substrates such as phenol and aniline derivatives has been investigated.

Conventional phenol resins (novolaks and resols) are widely used in industrial fields. However, the concern over the toxicity of formaldehyde monomer has resulted in limitations on their production and use. Recently, enzymatic synthesis of polyphenols has been extensively investigated since this methodology is expected to be an alternative process for preparation of phenolic polymers.

We have first investigated enzymatic polymerization of phenol catalyzed by horseradish peroxidase (HRP) (*21,22*). The polymerization was performed in a mixture of 1,4-dioxane and phosphate buffer (pH 7) (80:20 vol%). The polymerization started by the addition of hydrogen peroxide. The resulting polymer was powdery and partly soluble in DMF and DMSO, insoluble in methanol, chloroform, benzene, and water. From IR and NMR analysis, the polymer was found to be composed of a mixture of phenylene and oxyphenylene units (Figure 7). Solvent compositions such

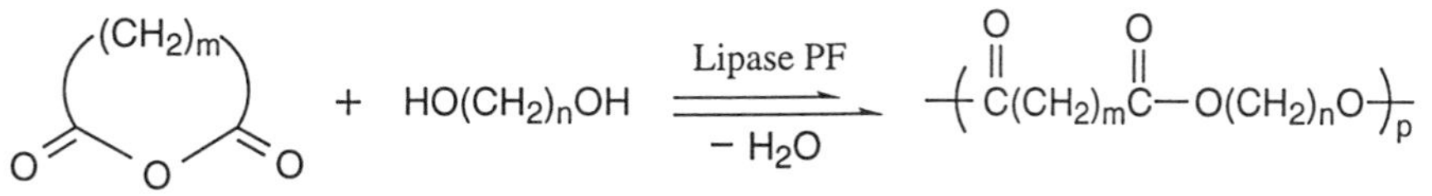

Figure 6. Lipase-catalyzed poly(addition-condensation) of cyclic acid anhydride with glycols.

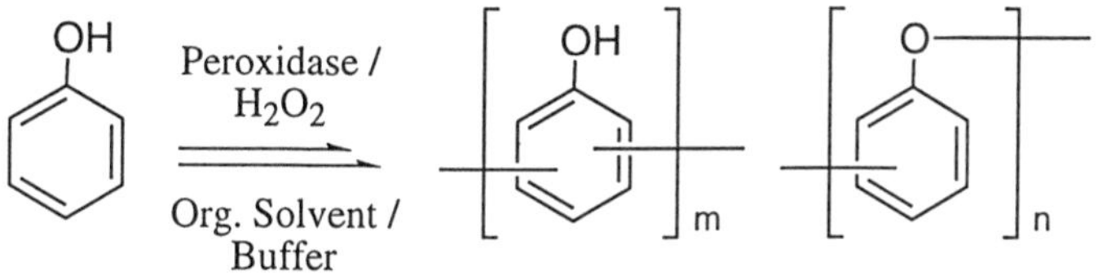

Figure 7. Peroxidase-catalyzed oxidative polymerization of phenol.

as type of organic solvent, buffer pH, and the mixed ratio greatly affected the yield, molecular weight, and solubility of the polymer (Table V). Soybean peroxidase (SBP) could catalyze the oxidative polymerization of phenol (23).

Table V. Enzymatic Oxidative Polymerization of Phenol[a]

	Polymerization			Polymer		
			Buffer			
Catalyst[b]	Org. Solvent	pH	Salt	Yield[c] (%)	DSP[d] (%)	Mn[e] (x10^{-4})
HRP	1,4-Dioxane	5	Acetate	80	24	3.2
HRP	1,4-Dioxane	7	Phosphate	75	25	2.5
HRP	1,4-Dioxane	7	Sulfite	0	---	---
HRP	1,4-Dioxane	8	Phosphate	64	31	3.6
HRP	1,4-Dioxane	10	Carbonate	20	12	1.9
HRP	Methanol	7	Phosphate	38	100	2.0
HRP	Acetone	7	Phosphate	71	11	>40 [f] 7.4 [f]
HRP	Acetonitrile	7	Phosphate	38	31	37 [f] 8.4 [f]
SBP	1,4-Dioxane	7	Phosphate	42	31	1.9

[a] Polymerization was carried out in a mixture of organic solvent and buffer (80:20 vol%) at room temperature for 24 h. [b] HRP: horseradish peroxidase; SBP: soybean peroxidase. [c] Isolated yield. [d] DMF-soluble part of the polymer. [e] Determined by GPC. [f] Bimodal peaks.

HRP-catalyzed polymerization of alkylphenols was examined in the aqueous 1,4-dioxane and reverse micellar system. In case of the polymerization of p-n-alkylphenols, the yield of the polymer increased with increasing the chain length from 1 to 5, and the yield of the polymer from hexyl or heptyl was almost the same as that from the pentyl derivatives (Figure 8) (24). On the other hand, these was a maximum point of the polymer yield in the reverse micellar system: the highest yield was obtained from p-ethylphenol. In case of isopropylphenols, polymer formation was observed from only p-isomer, whereas all cresol isomers were enzymatically polymerized (25). These results indicate that the polymerization behavior much depended on the substituent and substituted position.

Bisphenol derivatives, 4,4'-biphenol, bisphenol-A, 4,4'-methylenebisphenol, 4,4'-dihydroxydiphenyl ether, and 4,4'-thiodiphenol, were enzymatically polymerized to produce a new class of polyphenols . The HRP-catalyzed polymerization of 4,4'-biphenol in a mixture of 1,4-dioxane and phosphate buffer afforded a soluble polymer with molecular weight of several thousands (26). An oligomer showing high solubility toward polar organic solvents was obtained from bisphenol-A in a mixed solvent of acetone and phosphate buffer (20:80 vol%).

Thermal stability of the enzymatically synthesized polyphenols was evaluated by TG analysis. Figure 9 shows TG chart of polyphenol obtained by using HRP catalyst in the aqueous 1,4-dioxane. In the measurement under nitrogen, 10 % of the weight loss was observed at 387 ℃ and 43% of the polymer remained at 1000 ℃. The residue was thought to be carbonized polymer. Under air, the polymer weight was 10 % lost at 335 ℃ and completely decomposed at 571 ℃.

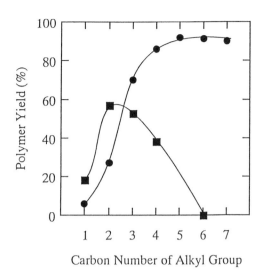

Carbon Number of Alkyl Group

Figure 8. Relationship between alkyl chain length of monomer and polymer yield in the HRP-catalyzed polymerization of *p-n*-alkylphenols (●) in a mixture of 1,4-dioxane and phosphate buffer (pH 7) and (■) in a reverse micellar solution.

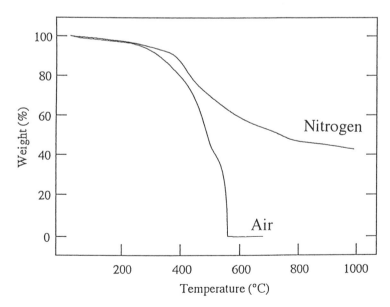

Figure 9. TG traces of polyphenol obtained by HRP catalyst in a mixture of 1,4-dioxane and phosphate buffer (pH 7) (80:20 vol%).

TG analysis data are summarized in Table VI. Thermal stability depended on the monomer structure as well as the polymerization catalyst. The polymer from 4,4'-biphenol had the highest temperature at 10 % loss and yield of the residue at 1000 ℃ under nitroten.

Table VI. TG Analysis of Polyphenol

Polymerization[a]			Polymer		
Monomer	Catalyst[b]	Atmosphere	T_{d10}[c] (℃)	T_{d100}[d] (℃)	Residue[e] (%)
Phenol	HRP	Air	335	571	---
Phenol	HRP	Nitrogen	387	---	43
Phenol	SBP	Air	277	523	---
Phenol	SBP	Nitrogen	305	---	37
p-Cresol	HRP	Air	263	544	---
p-Cresol	HRP	Nitrogen	277	---	40
p-n-Butylphenol	HRP	Nitrogen	288	---	26
p-n-Heptylphenol	HRP	Nitrogen	307	---	16
4,4'-Biphenol [f]	HRP	Air	360	548	---
4,4'-Biphenol [f]	HRP	Nitrogen	404	---	61
Bisphenol-A [f]	HRP	Nitrogen	294	---	32

[a] Polymerization was carried out in a mixture of 1,4-dioxane and phosphate buffer (pH 7) (80:20 vol%) at room temperature for 24 h. [b] HRP: horseradish peroxidase; SBP: soybean peroxidase. [c] Temperature at 10 weight % loss. [d] Temperature at complete decomposition. [e] Weight % of residue at 1000 ℃. [f] Data of methylated polymer.

Enzymatic Synthesis of Poly(phenylene oxide). Poly(1,4-oxyphenylene) (poly(phenylene oxide), PPO) is one of the common engineering plastics. It is prepared industrially by oxidative polymerization of 2,6-dimethylphenol using copper/amine catalyst system. We have achieved enzymatic synthesis of PPO from 3,5-dimethoxy-4-hydroxybenzoic acid (syringic acid) (Figure 10) (27). The polymerization is a new type of the enzymatic polymerization involving elimination of carbon dioxide and hydrogen from the monomer. Peroxidases (HRP and SBP) as well as laccase derived from *Pycnoporus coccineous* were effective as catalysts for the polymerization. The laccase-catalyzed polymerization in a mixture of acetone, chloroform, and acetate buffer afforded PPO with molecular weight of 1.8×10^4. PPO was also obtained by enzymatic polymerization of 2,6-dimethylphenol catalyzed by peroxidase or laccase.

HRP catalyzed oxidative polymerization of p-alkoxyphenols in the aqueous organic solvent to give the polymer soluble in chloroform, DMSO, and DMF. The polymer structure was found to be exclusively of phenylene oxide unit from NMR and IR analyses. These data indicate that the polymerization of p-alkoxyphenols proceeded regioselectively, leading to the formation of poly(phenylene oxide).

Preparation of Polyphenol Particles. Preparation of polyphenol particles was examined by HRP-catalyzed dispersion polymerization in a mixture of 1,4-dioxane and phosphate buffer (60:40 vol%). The polymerization of phenol using poly(vinyl methyl ether) as steric stabilizer afforded the polymer particles quantitatively, which was relatively monodisperse in the sub-micron range (*28*). Particle size could be controlled by the solvent composition or stabilizer concentration. The particles were also obtained from *p*-phenylphenol, and *o*- and *m*-cresols.

Enzymatic Polymerization of Aniline Derivatives. HRP catalysis induced the oxidative polymerization of aniline derivatives, yielding polyaromatics. The enzymatic polymerization of *o*-phenylenediamine in the aqueous 1,4-dioxane produced soluble poly(imino-2-aminophenylene) with molecular weight of 2×10^4 (Figure 11) (*29*), which can not be obtained by conventional oxidative polymerizations. Polymeric materials were enzymatically obtained from various *p*-substituted-*o*-phenylenediamines and *p*-phenylenediaine, whereas no polymer formation was observed from *m*-isomer.

HRP-catalyzed copolymerization of *o*-phenylenediamine with phenol produced polymeric materials (*30*). The copolymer was partly soluble in DMF and DMSO, but insoluble in other common solvents. IR analysis showed that the copolymer is composed of a mixture of the units obtained by the homopolymerization of the both monomers; phenylene and oxyphenylene units from phenol and iminophenylene unit from *o*-phenylenediamine.

Conclusion

New polymerization reactions catalyzed by hydrolase (lipase) and oxidoreductases (peroxidases and laccase) afforded aliphatic polyesters and polyaromatics, respectively. Lipase catalyzed ring-opening polymerization of lactones in various ring-sizes, poly(addition-condensation) of cyclic acid anhydrides and α,ω-glycols, and polycondensation of bis(enol ester)s and glycols to give the polyesters under mild reaction conditions. Single-step synthesis of end-functionalized polyesters, macromomer and telechelics, was achieved by the polymerization in the presence of vinyl esters. Oxidative polymerization of phenol and aniline derivatives by using oxidoreductase catalyst produced novel polyaromatics. The enzymatically synthesized polyphenol showed high thermal stability. Monodisperse polyphenol particles in the sub-micron range were produced by the dispersion polymerization using poly(vinyl methyl ether) as stabilizer.

In recent years, structural variation of synthetic targets on polymers has begun to develop highly selective polymerizations for the increasing demands in the production of various functional polymers in material science. Enzymatic polymerization is expected to create a new area of polymer synthesis where various

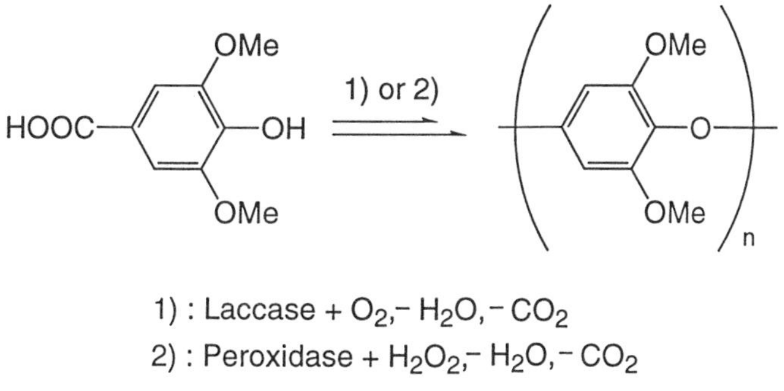

1) : Laccase + O_2,– H_2O, – CO_2
2) : Peroxidase + H_2O_2,– H_2O, – CO_2

Figure 10. Enzymatic oxidative polymerization of syringic acid.

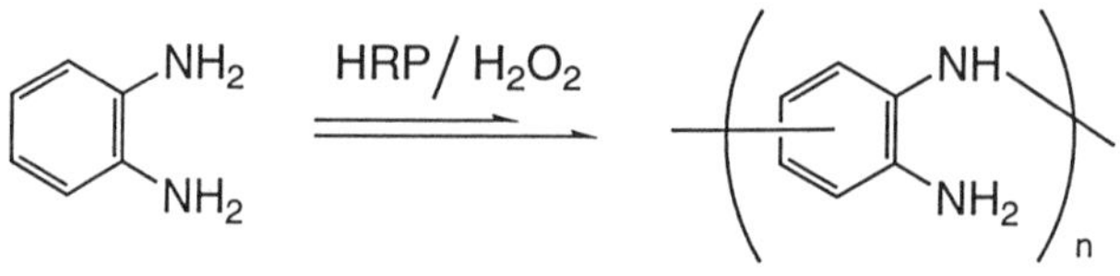

Figure 11. Peroxidase-catalyzed oxidative polymerization of o-phenylenediamine.

macromolecular materials, which have been difficult to produce by conventional methodologies, will be produced efficiently.

Acknowledgements

This work was supported by a Grant-in-Aid for Specially Promoted Research (No. 08102002) from the Ministry of Education, Science, and Culture, Japan.

Literatures Cited

(1)   Jones, J. B. *Tetrahedron* **1986**, *42*, 3351.
(2)   Klibanov, A. M. *Acc. Chem. Res.* **1990**, *23*, 114.
(3)   Santaniello, E.; Ferraboschi, P.; Grisenti, P.; Manzocchi, A. *Chem. Rev.* **1992**, *92*, 1071.
(4)   Kobayashi, S.; Shoda, S.; Uyama, H. *Adv. Polym. Sci.* **1995**, *121*, 1.
(5)   Ritter, H. *Trends Polym. Sci.* **1993**, *1*, 171.
(6)   Kobayashi, S.; Kashiwa, K.; Kawasaki, T.; Shoda, S. *J. Am. Chem. Soc.* **1991**, *113*, 3079.
(7)   Kobayashi, S.; Shimada, J.; Kashiwa, K.; Shoda, S. *Macromolecules* **1992**, *25*, 3237.
(8)   Shoda, S.; Okamoto, E.; Kiyosada, T.; Kobayashi, S. *Macromol. Rapid Commun.* **1994**, *15*, 751.
(9)   Namekawa, S.; Uyama, H.; Kobayashi, S. *Polym. J.* **1996**, *28*, 730.
(10)   Uyama, H.; Kobayashi, S. *Chem. Lett.* **1993**, 1149.
(11)   Uyama, H.; Takeya, K.; Hoshi, N.; Kobayashi, S. *Macromolecules* **1995**, *28*, 7046.
(12)   Uyama, H.; Takeya, K.; Kobayashi, S. *Bull. Chem. Soc. Jpn.* **1995**, *68*, 56.
(13)   Uyama, H.; Kikuchi, H.; Takeya, K.; Kobayashi, S. *Acta Polymerica* **1996**, *47*, 357.
(14)   Uyama, H.; Takeya, K.; Kobayashi, S. *Proc Jpn. Acad.* **1993**, *69B*, 203.
(15)   Uyama, H.; Namekawa, S.; Kobayashi, S. *Polym J.* **1997**, *29*, 299.
(16)   Uyama, H.; Kikuchi, H.; Kobayashi, S. *Chem. Lett.* **1995**, 1047.
(17)   Uyama, H.; Kobayashi, S. *Chem. Lett.* **1994**, 1687.
(18)   Namekawa, S.; Uyama, H.; Kobayashi, S. *Polym. Prepr., Jpn.* **1996**, *45*, 200.
(19)   Kobayashi, S.; Uyama, H. *Makromol. Chem., Rapid Commun.* **1993**, *14*, 841.
(20)   Uyama, H.; Kikuchi, H.; Takeya, K.; Hoshi, N.; Kobayashi, S. *Chem. Lett.* **1996**, 107.
(21)   Uyama, H.; Kurioka, H.; Kaneko, I.; Kobayashi, S. *Chem. Lett.* **1994**, 423.

(22)   Uyama, H.; Kurioka, H.; Sugihara, J.; Kobayashi, S. *Bull Chem Soc. Jpn.*, **1996**, *69*, 189.

(23)   Uyama, H.; Kurioka, H.; Komatsu, I.; Sugihara, J.; Kobayashi, S. *Macromol. Reports* **1995**, *A32*, 649.

(24)   Kurioka, H.; Komatsu, I.; Uyama, H.; Kobayashi, S. *Macromol. Rapid Commun.* **1994**, *15*, 507.

(25)   Uyama, H.; Kurioka, H.; Sugihara, J.; Komatsu, I.; Kobayashi, S. *Bull Chem Soc. Jpn.*, **1995**, *68*, 3209.

(26)   Kobayashi, S.; Kurioka, H.; Uyama, H. *Macromol. Rapid Commun.* **1996**, *17*, 503.

(27)   Ikeda, R.; Uyama, H.; Kobayashi, S. *Macromolecules* **1996**, *29*, 3053.

(28)   Uyama, H.; Kurioka, H.; Kobayashi, S. *Chem. Lett.* **1995**, 795.

(29)   Kobayashi, S.; Kaneko, I.; Uyama, H. *Chem. Lett.* **1992**, 393.

(30)   Uyama, H.; Kurioka, H.; Kaneko, I.; Kobayashi, S. *Macromol. Reports* **1994**, *A31*, 507.

Chapter 4

Enzyme-Catalyzed Ring-Opening Polymerization of Four-Membered Lactones

Preparation of Poly(β-propiolactone) and Poly(β-malic acid)

Shuichi Matsumura, Hideki Beppu, and Kazunobu Toshima

Department of Applied Chemistry, Faculty of Science and Technology, Keio University, 3-14-1, Hiyoshi, Kohoku-ku, Yokohama 223, Japan

Lipase-catalyzed ring-opening polymerization of β-propiolactone (PL) and benzyl β-malolactonate (BM) was studied with respect to the influence of reaction conditions on molecular weight of the polymers and the monomer conversion. PL and BM were polymerized in bulk using lipase as a catalyst to yield the corresponding polyesters having molecular weights of greater than 50,000 and 7,000, respectively. It was found that the molecular weight of poly(β-propiolactone) produced by the lipase-catalyzed polymerization of PL was inversely dependent on the concentration of lipase. Both molecular weight and polymerization speed were increased with increasing reaction temperature from 40 to 60°C. BM was readily polymerized by porcine pancreatic lipase or Novozym 435 lipase at 60°C to yield poly(benzyl β-malate), which was readily debenzylated by catalytic hydrogenolysis to form poly(β-malic acid). This polyester was readily biodegraded both by the aerobic and anaerobic microbes.

Much interest in polyesters such as poly(glycolic acid), poly(malic acid), and poly(β-propiolactone) [poly(PL)] is mainly due to their expanding use as degradable materials in medicinal and pharmaceutical applications (1-8). It is known that the four-membered lactone, β-propiolactone (PL), was polymerized by both anionic and cationic polymerization to produce a high-molecular weight poly(PL) (9,10) and by living polymerization (11-13). Poly(PL) was also prepared by radiation-induced polymerization of PL at low temperature, and the in vivo degradation was analyzed (14). Hydrolytic degradation of melt-extracted fibers from poly(PL) was analyzed, and the degradation was found to be very rapid during the first 90 days of immersion in a buffered salt solution at pH 7.2. (1).

High-molecular weight polycarboxylates have been shown to have excellent properties both in the industrial and biomedical fields. However, they are generally highly resistant to biodegradation which is an important criterion in large scale applications. Among the high-molecular weight polycarboxylates, only a few polymers which contain ester or amide linkages in the backbone, such as poly(malic acid) (15,16), poly(γ-glutamic acid) (17,18) and poly(aspartic acid) (19-23), are biodegradable (*24,25*). Poly(malic acid) is a biodegradable and bioadsorbable water-soluble polyester having modifiable pendant carboxylic groups. Recently, this polymer attracted attention as a polymer carrier which is able to covalently attach drug units and targeting agents in the pharmaceutical fields (*4-7*) and can also be used as a biodegradable raw material for the chemical industries such as detergent builders and chelating agents (*15*). Natural poly(malic acid) has been reported to be produced by *Penicillium cyclopium* as a protease inhibitor (*26,27*) and by *Physarum polycephalum* as a DNA polymerase inhibitor (*28*). Recently, fermentation production of poly(β-L-malic acid) by *Aureobasidium* sp. was reported (*29*). However, these fermentation processes may not be feasible for the preparation of molecular designed poly(malic acid) copolymers. The chemical method for the preparation of poly(β-malic acid), first reported by Vert and Lenz, showed that the ring-opening polymerization of benzyl β-malolactonate is involved (*30-33*). Optically active poly(β-malic acid) with high optical purity was prepared by the anionic polymerization of optically active benzyl malolactonate which was prepared from L-aspartic acid. Thus obtained isotactic optically active poly(benzyl β-malate) was selectively debenzylated to yield optically active poly(β-malic acid) (*32*). Furthermore, copolymerization by the combination of benzyl β-malolactonate and other lactones were reported (*34-36*)

The ring-opening polymerization of the lactone requires extremely pure monomers and anhydrous conditions as well as a long reaction time. Furthermore, the polymerization catalyst may be present in the resultant polymer, and additional purification procedures will be needed for medical applications. To avoid these difficult restrictions for ring-opening polymerization of lactones by the chemical methods, enzyme-catalyzed polymerization may be one of the feasible methods to obtain polyesters.

Lipase-catalyzed ring-opening polymerization of six and seven-membered lactones was first conducted using lipase as a catalyst by Uyama *et. al.* (*37*). In recent years, enzymatic polymerization has been expanded to the ring-opening polymerization and copolymerization of medium-size lactones and macrolides (12-, 13- and 16-membered) yielding various polyesters (*38,39*). They reported that the enzymatic polymerizability showed a reverse direction of the ring-strain (*40,41*). An extensive study of the enzyme-catalyzed ring-opening polymerization of ε-caprolactone was also published (*42*). The ring-opening polymerization using lipase, so far reported, were restricted to lactones greater than 6-membered lactones. Very recently, the enzymatic polymerization of the four-membered lactones, PL (*43,44*), β-butyrolactone (*45*) and α-methyl-β-propiolactone (*46*) has been reported. However, enzymatic ring-opening polymerization of benzyl β-malolactonate has not been reported yet.

In this report, lipase-catalyzed ring-opening polymerization of PL (*43*) and benzyl β-malolactonate (*16,47*) was studied with respect to the influence of reaction conditions on molecular weight of the polymers and the monomer conversion.

Experimental

Materials and measurements. β-Propiolactone (PL) was purchased from Tokyo Kasei Kogyo Co., Ltd. (Tokyo, Japan) and dried over Na-Pb (Dry-soda, Nacalai Tesque, Inc., Kyoto, Japan) and then distilled under reduced pressure (36°C/5 torr) in a nitrogen atmosphere before use for polymerization. Porcine pancreatic lipase (41 U/mg protein, according to the supplier) and lipase from *Candida cylindracea* (500 U/mg, according to the supplier) were purchased from Sigma Chemical Co. (St. Louis, MO, USA). Novozym 435 (triacylglycerol hydrolase + carboxyesterase) having 7,000 PLU/g (propyl laurate units) and Lipozyme IM (triacylglycerol hydrolase) were kindly supplied by Novo Nordisk A/S (Bagsvaerd, Denmark).

Lipase PS was kindly supplied by Amano Pharmaceutical Co., Ltd. The enzymes were used without further purification. The other materials used were of the highest available purity.

Benzyl β-D,L-malolactonate (BM) was prepared according to the chemical method of Lenz and Vert et al. (30-33,15) as shown in scheme 1. Maleic anhydride was treated with hydrogen bromide to yield bromosuccinic anhydride. This compound was converted to the monobenzyl ester of bromosuccinic acid by treatment with benzyl alcohol in the presence of concentrated sulfuric acid as a catalyst. In this reaction sequence, as noted in the scheme, two different half-esters were obtained with R1 and R2 being a either H or a benzyl group. The desired intermediate has the benzyl group at the C1 carboxyl group. This compound was converted to BM by treatment with sodium carbonate.

The number-average molecular weight (M_n), weight-average molecular weight (M_w) and molecular weight dispersion (M_w/M_n) were measured by gel-permeation chromatography (GPC) using GPC columns (Shodex 80M, Showa Denko Co., Ltd., Tokyo, Japan) with a reflective index detector. Chloroform was used as the eluent. The GPC system was calibrated with polystyrene standards. ^{1}H-NMR spectra were recorded with a JEOL Model GSX-270 (270 MHz) spectrometer. ^{13}C-NMR spectra were determined with a JEOL model JNM-FX90A Fourier transform spectrometer, operating at 22.5 MHz with complete proton decoupling (JEOL Ltd., Tokyo, Japan).

Polymerization procedure. Lipase-catalyzed ring-opening polymerization of PL and BM was carried out as shown in scheme 2. A mixture of lipase and lactone with/without organic solvent in argon atmosphere was kept in a sealed glass tube equipped with a magnetic stirring bar in a thermostated oil bath. Magnetic stirring was used for suspension of the enzyme in a monomer in the initial step of polymerization. After the reaction, the reaction mixture was dissolved in chloroform, and the insoluble enzyme was removed by filtration; then the filtrate was evaporated under reduced pressure to obtain the polymer. The molecular weight of the polymer was analyzed by GPC. The conversion of PL to poly(PL) was determined by comparison of the spectral integration intensities of the triplet at 3.5 ppm corresponding to the methylene group adjacent to the lactone-ring oxygen of PL with respect to the poly(PL) intrachain protons at 2.6 ppm. The conversion of BM to poly(benzyl β-D,L-malolactonate) [poly(BM)] was analyzed by GPC. The polymer was further purified by reprecipitation to remove any unreacted monomeric lactone and oligomers (chloroform as a good solvent; methanol as a poor solvent). The chemical polymerization of BM was carried out in a similar way except that betaine was used in place of the enzyme as the polymerization catalyst. The molecular structure was analyzed by FT-IR, ^{1}H-NMR and ^{13}C-NMR spectroscopy and elemental analysis.

Results and Discussion

Lipase-catalyzed polymerization of PL. PL was found to be readily polymerized by the lipase to yield poly(PL). Figure 1 shows the GPC profile changes of poly(PL) during the enzyme-catalyzed polymerization of PL using 0.5% *Candida cylindracea* lipase at 60°C and 80°C. It was observed that the peak top of the GPC was gradually shifted towards high-molecular weight and the polymer peak area was gradually enlarged with increasing monomer conversion. It was also observed from the GPC analysis that the polymerization rate of PL at 80°C was much faster than that at 60°C. Typical polymerization conditions and analytical data are shown in Table I. Polymerization of PL occurred with all lipases tested in this experiment. However, significant differences in the polymerization results between the enzymes were observed with respect to the conversion, the molecular weight and the molecular

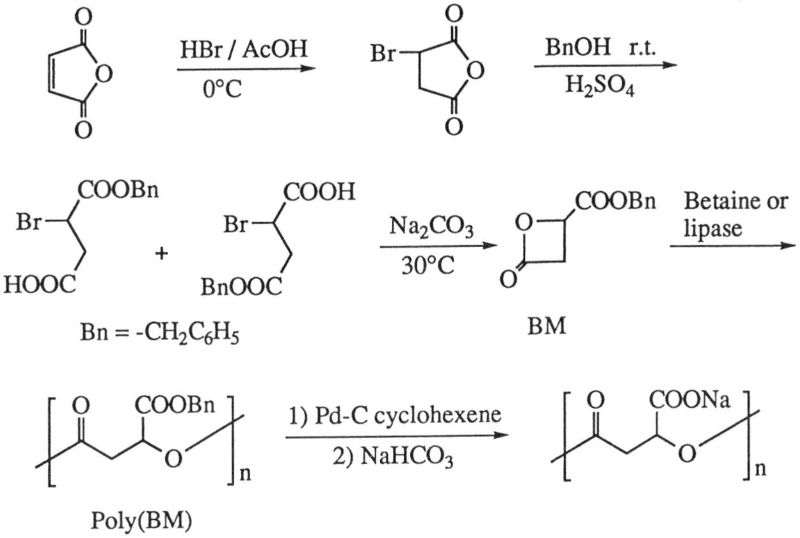

Scheme 1 Preparation of Poly (sodium β-D,L-malate)

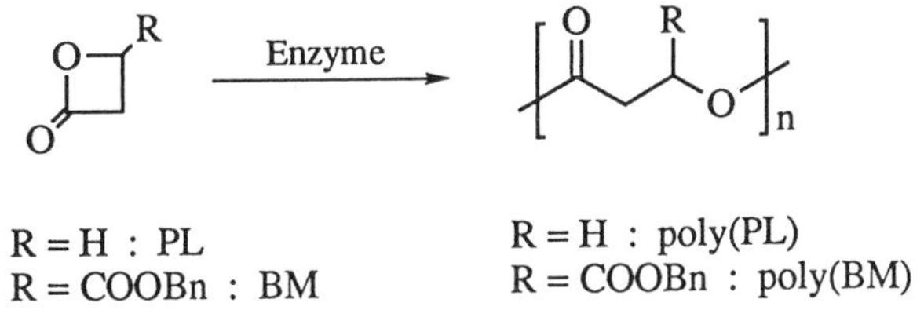

R = H : PL
R = COOBn : BM

R = H : poly(PL)
R = COOBn : poly(BM)

Scheme 2 Enzyme-Catalyzed Polymerization of β-Lactones

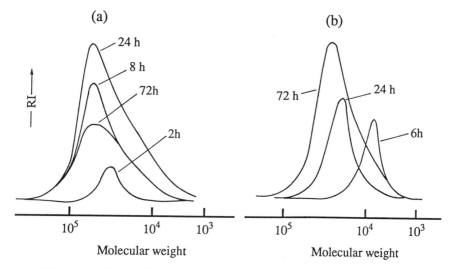

Figure 1.   GPC profile changes of poly(PL) during the enzyme-catalyzed
polymerization of PL using 0.5% *Candida cylindracea* lipase.
(a) : 80°C, (b) : 60°C.
(Adapted from ref. 43.)

Table I. Typical Lipase-Catalyzed Ring-Opening Polymerization
of β-Propiolactone (PL)

Entry[a]	Lipase[b]	wt %	Temp. (°C)	Time (h)	Conv. (%)	M_w	M_w/M_n
1	PPL	1.0	80	48	98	33800	2.9
2	CC	1.0	80	48	>99	35200	2.4
3	Novo	1.0	80	48	74	17800	6.7
4	---	---	80	48	50	7500	6.7
5	PPL	1.0	60	48	96	36500	3.7
6	CC	1.0	60	48	95	42700	2.1
7	CC	0.5	60	48	99	49100	2.2
8	Novo	1.0	60	48	32	5000	6.9
9	Lipo	1.0	60	48	>99	11000	4.2
10	PS	1.0	60	48	54	55000	1.6
11	---	---	60	48	13	5200	1.3
12	PPL	1.0	40	48	24	20400	1.5
13	CC	1.0	40	48	13	15400	1.7
14	Novo	1.0	40	48	4	---	---
15	---	---	40	48	3	---	---

[a] Entries 4, 11, 15 : blank tests. [b] PPL: porcine pancreatic lipase, CC: *Candida cylindracea* lipase, Novo: Novozym 435, Lipo: Lipozyme IM, PS: lipase PS. (SOURCE: Adapted from ref. 43.)

weight dispersion of the polymers. Among the lipases tested, porcine pancreatic lipase, *Candida cylindracea* lipase, and lipase PS showed the highest activities. When compared with the polymers obtained after 48-hour polymerization, both conversion and molecular weight of the polymers obtained at 60 and 80°C were higher than those obtained at 40°C. At 80°C, PL was found to be thermally oligomerized to M_n=1,100 or M_w=7,500 at 50% monomer conversion (entry 4 in Table I). However, the molecular weight of the thermally produced oligomer was significantly small comparing to that obtained by the lipase-catalyzed polymerization. On the other hand, thermal oligomerization of PL at 60 and 40°C was very low (entries 11 and 15 in Table I). These results indicate that the polymerization reaction of PL occurred mainly due to the lipase.

After purification by reprecipitation for further analysis of the polymer structure, FT-IR and ^{1}H-NMR of the purified poly(PL) was analyzed. The spectral data and elemental analysis of poly(PL) having an M_w of 49,100 are shown (entry 7 in Table I) as representative. IR(KBr): 2928, 1460 (CH$_2$), 1732, 1273 1172 (ester C=O) cm^{-1}. ^{1}H-NMR (CDCl$_3$): δ=2.65 (t; 2H, 6.0), 4.37 (t, 2H, 6.0). ^{13}C-NMR (CDCl$_3$): δ=33.7 ($\underline{C}$H$_2$C=O), 60.1 ($\underline{C}$H$_2$O), 170.3 (C=O). Elemental analysis, Found: C, 49.73; H, 5.77%. Calcd for (C$_3$H$_4$O$_2$)$_n$: C, 50.00; H, 5.60%.

Figure 2 shows the M_w and M_n of [poly(PL)] as a function of monomer conversion at reaction temperatures of 40, 60 and 80°C, respectively. At 80°C, the M_n of the polymer quickly reached a maximum value of 30,000. At this point, the reaction mixture was still fluid at 80 °C, and the polymerization might be carried out like a solution polymerization. The narrow molecular weight dispersion (M_w/M_n) at this conversion (20% conversion in Figure 2a) is probably ascribed to the mobility of enzyme and the reactant species in the reaction mixture. The reaction mixture then became very viscous or almost solid; further polymerization and depolymerization reactions still occurred slowly with increasing M_w, thus broadening the M_w/M_n. Similar tendencies were observed at 60°C. Because the M_n was considerably influenced by the low-molecular weight fractions in the polymer and did not clearly reflect the molecular weight increase due to a condensation-type reaction, in this report M_w was used to compare the molecular weight of the polymer.

Time course of lipase-catalyzed polymerization of PL. Figure 3 shows the relationship between reaction time and molecular weight as well as monomer conversion in the lipase-catalyzed polymerization of PL by *Candida cylindracea* lipase at an enzyme concentration of 0.5% at 80, 60 and 40°C, respectively. At 80°C, both the conversion and the M_w quickly increased within the first 10 hours; then the M_w of the polymer was slightly decreased, but the conversion was gradually increased to over 99%. At 60°C, the polymerization speed was decreased, and the M_w of the polymer reached the highest value after 48 hours' reaction; the M_w of the polymer then remained the same. The conversion was gradually increased to 95% after 72 hours. Decrease in the polymerization and monomer conversion rate after 48 hours is ascribed to the solidification of the polymerization system. On the other hand, the polymerization reaction was considerably decreased at 40°C, and the M_w did not reach the maximum value after 100 hours reaction; the monomer conversion remained about 20% after 50 hours reaction. The polymerization using 0.5% Lipase PS at 60°C showed tendencies similar to that using *Candida cylindracea* lipase at an enzyme concentration of 0.5% at 60°C. The M_w of the polymer was more quickly increased to 57,000 after 20 hours at 60°C compared to that using *Candida cylindracea* lipase at the same temperature. However, the rate of monomer conversion was slower than that using *Candida cylindracea* lipase at 60°C.

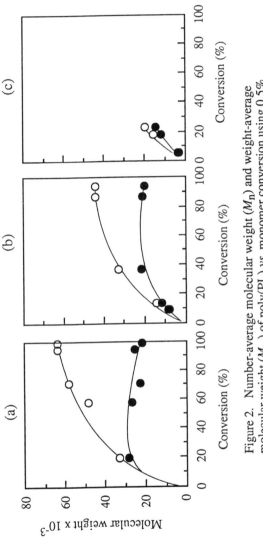

Figure 2. Number-average molecular weight (M_n) and weight-average molecular weight (M_w) of poly(PL) vs. monomer conversion using 0.5% *Candia cylindracea* lipase.
(a) : 80°C, (b) : 60°C, (c) : 40°C.   ● : M_n,   ○ : M_w
(Adapted from ref. 43.)

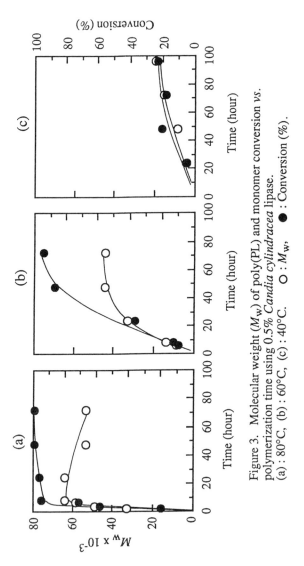

Figure 3. Molecular weight (M_w) of poly(PL) and monomer conversion vs. polymerization time using 0.5% *Candia cylindracea* lipase.
(a) : 80°C, (b) : 60°C, (c) : 40°C.   O : M_w,   ● : Conversion (%).

Influence of enzyme concentration upon the molecular weight of the polymer. It was found that the M_W of poly(PL) obtained by the lipase-catalyzed polymerization of PL inversely depends on the enzyme concentration in the reaction. Figure 4 shows the relationship between the M_W of the poly(PL) after 48 hours' reaction and the enzyme concentration in the polymerization reaction using *Candida cylindracea* lipase at 80, 60 and 40°C, respectively. The M_W of poly(PL) rapidly decreased with increasing enzyme concentration from 0.5% to 20%. It seemed that the most preferable reaction conditions for PL polymerization by *Candida cylindracea* lipase were 60°C and 0.5% lipase concentration with respect to M_W.

Lipase-catalyzed polymerization of BM. It was found that BM was readily polymerized by the lipase to yield poly(BM) with a M_W of greater than 7,000. After purification by reprecipitation for further analysis of the polymer structure, FT-IR, ^{1}H-NMR and ^{13}C-NMR of the purified polymers were analyzed. The spectral data and elemental analysis of poly(BM) having an M_W of 7,300 are shown (entry 6 in Table II) as representative. IR(KBr): 2957 (CH_2), 1748, 1163, 1055 (ester C=O), 1456 (CH_2), 752, 689 (aromatics) cm[-1]. ^{1}H-NMR (CDCl_3): δ=2.9 (2H, CH_2), 5.1 (2H, PhCH_2), 5.5 (1H, CH), 7.3 (5H, Ph). ^{13}C-NMR (CDCl_3): δ=35.4 (C̲H_2C=O), 67.6 (CH), 68.6 (PhC̲H_2), 128.2, 128.5, 128.6, 135.0 (aromatics), 168.0, 168.1 (C=O). Elemental analysis, Found: C, 64.00; H, 4.81%. Calcd for (C_{11}H_{10}O_4)n : C, 64.07; H, 4.89%. ^{1}H-NMR, ^{13}C NMR and IR spectra of the poly(BM) obtained in this report completely agreed with those of an authentic sample (*15*). Table II shows the typical ring-opening polymerization of BM by both the enzymatic method and the chemical method. It was confirmed that polymerization occurred with porcine pancreatic lipase and microbial origin lipase, Novozym 435. No significant difference between the two enzymes, porcine pancreatic lipase and microbial origin lipase, was observed with respect to the molecular weight of the polymer. However, both the monomer conversion and the yield of poly(BM) by the polymerization of BM using porcine pancreatic lipase were higher than that using Novozym 435. During the lipase-catalyzed polymerization of BM using porcine pancreatic lipase and Novozym 435, no significant formation of BM oligomer was detected by GPC. It was also confirmed that BM was not polymerized to produce poly(BM) without lipase under these conditions (entry 15 in Table II), indicating that the lipase catalyzed the polymerization of BM.

It was reported by Guerin *et al.* that the optical rotation of optically active (-)-poly(BM) with a molecular weight of 6,000, which was obtained from L-aspartic acid, was $[\alpha]_D^{25}$ -5.0° (c=1, CH_2Cl_2) (*32*). When compared to the optically active (-)-poly(BM), the optical activity of poly(BM) obtained by the lipase-catalyzed polymerization was low (entries 2 and 9 in Table II). However, poly(BM) obtained by Novozym 435 showed a slightly higher optical activity than that obtained by porcine pancreatic lipase. Further analysis is now under study.

The addition of an organic solvent, such as isooctane and heptane, did not significantly affect the molecular weight and molecular weight dispersions of the resultant polymer. When butanol was added to the polymerization mixture, it was confirmed by ^{1}H-NMR that the butyl moiety was incorporated into a terminal of the polymer chain, suggesting that poly(BM) had a linear structure (entries 11 and 12 in Table II). However, the molecular weight of poly(BM) decreased with the addition of butanol under these conditions.

Compared to the chemical polymerization of BM using betaine (entries 13 and 14 in Table II), the lipase-catalyzed ring-opening polymerization of BM tended to occur smoothly with respect to the reaction time and the molecular weight of the resultant polymer.

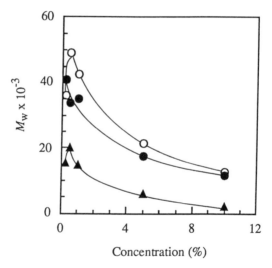

Figure 4. Molecular weight (M_w) of poly(PL) *vs.* enzyme concentration using
0.5% *Candida cylindracea* lipase after 48 hours reaction.
● : 80°C, ○ : 60°C, ▲ : 40°C.
(Reproduced with permission from ref. 43. Copyright 1996 Chapman & Hall.)

Table II. Typical Lipase-Catalyzed Ring-Opening
Polymerization of Benzyl β-D,L-Malolactonate (BM)[a]

Entry	Lipase[b]	%[c]	Solvent	Temp. (°C)	Time (d)	Yield (%)	M_w	M_w/M_n
1	PPL	10	---	60	3	60	5000	1.4
2	PPL	2.5	---	60	3	83	7000	2.0
3	Novo	10	---	60	3	17	7200	1.5
4	Novo	5	---	60	3	26	6500	1.9
5	Novo	2.5	---	60	3	26	6600	1.5
6	PPL	10	Isooctane	60	3	90	7300	1.8
7	PPL	5	Heptane	60	3	61	7200	1.9
8	Novo	10	Isooctane	60	3	17	4500	1.4
9	Novo	5	Isooctane	60	3	22	4200	1.4
10	Novo	5	Heptane	60	3	9	7200	1.6
11	PPL[d]	5	Isooctane	60	3	69	7200	1.8
12	Novo[d]	5	Heptane	60	3	13	5200	1.3
13	Betaine	0.12	---	40	19	30	4200	1.1
14	Betaine	0.03	---	40	29	42	4400	1.1
15	---	---	---	60	3	0	---	---

[a] Entry 2: $[\alpha]_D^{29}$ + 0.33 (c 0.92, $CHCl_3$), Entry 9: $[\alpha]_D^{30}$ + 1.80 (c 1.02, $CHCl_3$). [b] PPL : porcine pancreatic lipase, Novo: Novozym 435, betaine : N,N,N-trimethyl glycine. [c] Weight % of lipase to BM. [d] 0.062 M Butanol (final concentration) was added. (SOURCE: Adapted from ref. 47.)

Time course of lipase-catalyzed polymerization of BM. Figure 5 shows the relationship between reaction time and M_W as well as monomer conversion of BM by porcine pancreatic lipase at an enzyme concentration of 10% at 60°C. It was observed that the M_W of the resultant polymer reached the highest value after 24 h, and this M_W remained the same. On the other hand, the conversion was gradually increased with the reaction time. These results indicate that the polymerization by porcine pancreatic lipase will proceed like a chain polymerization.

Debenzylation of poly(BM). The benzyl group of poly(BM) was readily removed by catalytic hydrogenation using Pd/C in the presence of cyclohexene or hydrogen gas to yield poly(β-D,L-malic acid) almost quantitatively (*4,15,32*). This was then neutralized by aqueous sodium hydrogen carbonate to almost quantitatively obtain the corresponding poly(sodium β-malate).

Microbial degradability of poly(sodium β-malate). The biodegradability of the polymers can be predicted by measuring the BOD values. The BOD values were measured with a BOD tester, an activated sludge freshly obtained from a municipal sewage treatment plant, and a test polymer concentration of 25 ppm, basically according to the OECD Guidelines for Testing of Chemicals (301C, Modified MITI Test) at 25°C. As shown in Figure 6, the poly(sodium β-malate) was confirmed to show excellent biodegradability with activated sludge. The extent of BOD biodegradability exceeded 60% which is a criterion for readily biodegradation.
    Water-soluble polymeric compounds will be widely diffused into aerobic environments as well as anaerobic environments of soil, river water or sea water of the earth after their use. It will be very important to estimate the biodegradability of such water-soluble polymers under anaerobic conditions in addition to aerobic conditions. Anaerobic biodegradability of the polycarboxylates was demonstrated using anaerobic river sediments (*48-50*). Figure 6 shows the biodegradability as determined by the total organic carbon (TOC) value after biodegradation and the initial TOC value of the culturing media. It was confirmed that the biodegradability of poly(sodium β-malate) was comparable to that of D-glucose and more than 60% of the organic carbon was removed from the incubation media by the anaerobic degradation. Moreover, the rate of biodegradation was relatively comparable to that under aerobic conditions.

Conclusions

    It was found that four-membered lactones, PL and BM, were readily polymerized in bulk by lipase to yield the corresponding high molecular weight poly(PL) and poly(BM), respectively. The molecular weight of the polymer produced by the lipase-catalyzed polymerization of PL was inversely dependent on the concentration of lipase in the polymerization. Both molecular weight and polymerization speed of PL were increased with increasing temperature from 40 to 60°C. High-molecular weight poly(sodium β-D,L-malate) could be prepared by the ring-opening polymerization of BM using lipase followed by subsequent debenzylation. Poly(sodium β-D,L-malate) was readily biodegraded under both aerobic and anaerobic conditions.

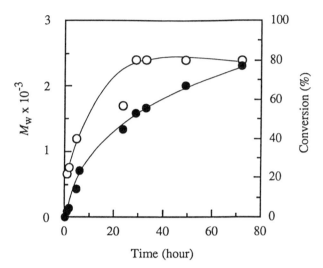

Figure 5. Molecular weight (M_w) of poly(BM) and monomer conversion *vs.* reaction time using 10% porcine pancreatic lipase at 60°C.
O : M_w,   ● : Conversion (%).

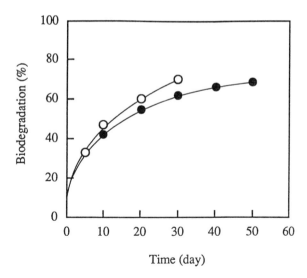

Figure 6. Biodegradation of poly(sodium β-D,L-malate) having M_w=6800 under aerobic and anaerobic conditions.
O : aerobic biodegradation,   ● : anaerobic biodegradation.

References

1. Mathisen, T.; Albertsson, A.-C. *J. Appl. Polym. Sci.* **1990**, *39*, 591.
2. Guerin, P.; Vert, M.; Braut, C.; Lenz, R.W. *Polym. Bull.* **1985**, *14*, 187.
3. Arnold, S.C.; Lenz, R.W. *Makromol. Chem., Macromol. Symp.* **1986**, *6*, 285.
4. Ouchi, T.; Fujino, A. *Makromol.Chem.* **1989**, *190*, 1523.
5. Braud, C.; Vert, M. *Polym. Prepr. (Am. Chem. Soc., Div. Polym. Chem.)* **1983**, *24*, 71.
6. Fourrnie, P.; Domurado, D.; Guerin, P.; Braud, C.; Vert, M.; Madelmont, J.-C. *J. Bioact. Compat. Polym.* **1990**, *5*, 381.
7. Ouchi, T.; Fujino, A.; Tanaka, K.; Banba, T. *J. Controlled Release* **1990**, *12*, 143.
8. Kimura, Y.; Shirotani, K.; Yamane, H.; Kitao, T. *Macromolecules* **1988**, *21*, 3340.
9. Johns, D.B.; Lenz, R.W.; Luecke, A. In *Ring-Opening Polymerization*; Iving, K.J.; T.Saegusa, Eds.; Elsevier, London, 1984, 461.
10. Slomkowski, S.; Penczeck, S. *Macromolecules* **1976**, *9*, 367.
11. Slomkowski, S.; Penczeck, S. *Macromolecules* **1983**, *13*, 229.
12. Yasuda, T.; Aida, T.; Inoue, S. *Macromolecules* **1983**, *16*, 1792.
13. Asano, S.; Aida, T.; Inoue, S. *Macromolecules* **1985**, *18*, 2057.
14. Asano, M.; Yoshida, M.; Kaetsu, I.; Morita, I.; Y., Fukuzaki, H.; Mashimo, T.; Yuasa, H.; Imai, K.; Yamanaka, H.; Kawaharada, U.; Suzuki, K. *Kobunshi Ronbunshu* **1987**, *44*, 897.
15. Abe, Y.; Matsumura, S.; Imai, K. *J. Jpn. Oil Chem. Soc.* **1986**, *35*, 937.
16. Matsumura, S.; Beppu, H.; Toshima, K. *PMSE (Am. Chem. Soc., Div. Polym. Materials: Sci. Eng.)* **1996**, *74*, 2.
17. Kunioka, M.; Goto, A. *Appl. Microbiol. Biotech.* **1994**, *40*, 867.
18. Cromwell, A.-M.; Gross, R.A. *Int. J. Biol. Macromol.* **1995**, *17*, 259.
19. Low, K.C.; Wheeler, A.P.; Koskan, L.P. In *Hydrophilic Polymers;* Glass, J.E., Ed.; Adv. Chem. Ser. 248; Am. Chem. Soc.: Washington, 1996, 99.
20. Alford, D.D.; Wheeler, A.P.; Pettigrew, C.A. *J. Environ. Polym. Degrad.* **1994**, *2*, 225.
21. Swift, G. *Accounts Chem. Res.* **1993**, *26*, 105.
22. Freeman, M.B.; Paik, Y.H.; Swift, G.; Wilczynski, R.; Wolk, S.K.; Yocom, K.M. In *Hydrogels and Biodegradable Polymers for Bioapplications;* Ottenbrite, R.M.; Huang, S.J.; K. Park, K., Eds.; ACS Symp. Ser. 627; Am. Chem. Soc.: Washington, 1996, 118.
23. Henderson, L.A.; Svirkin, Y.Y.; Gross, R.A.; Kaplan, D.L.; Swift, G. *PMSE (Am. Chem. Soc., Div. Polym. Materials: Sci. Eng.)* **1996**, *74*, 6.
24. Kawai, F. *Adv. Biochem. Eng./Biotech.* **1995**, *52*, 151.
25. Swift, G. *Polym. Degrad. Stabil.* **1994**, *45*, 215.
26. Shimada, K.; Matsushima, K. *Agric. Biol. Chem.* **1969**, *33*, 544.
27. Shimada, K.; Matsushima, K. *Agric. Biol. Chem.* **1969**, *33*, 549.
28. Fischer, H.; Erdmann, S.; Holler, E. *Biochemistry* **1989** *28*, 5219.
29. Nagata, N.; Nakahara, T.; Tabuchi, T. *Biosci. Biotech. Biochem.* **1993**, *57*, 638.
30. Vert, M.; R.W. Lenz, R.W. *Polym. Prepr. (Am. Chem. Soc., Div. Polym. Chem.)* **1979**, *20*, 608.
31. Lenz, R.W.; Vert, M. *U.S. Pat.* 4, 265,247 (May 5, 1981), 4,320,753 (March 23, 1982)
32. Guerin, P.; Vert, M.; Braud, C.; Lenz, R.W. *Polym. Bull.* **1985**, *14*, 187.
33. Arnold, S.C.; Lenz, R.W. *Makromol. Chem., Macromol. Symp.* **1986**, *6*, 285.
34. Otani, N.; Kimura, Y.; Kitao, T. *Kobunshi Ronbunshu* **1987**, *44*, 701.
35. Gross, R.A.; Zhang, Y.; Konrad, G.; Lenz, R.W. *Polym. Prepr. (Am. Chem. Soc., Div. Polym. Chem.)* **1987**, *28(2)*, 373.
36. Benevenuti, M.; Lenz, R.W. *J. Polym. Sci., Polym. Chem. Ed.* **1991**, *29*, 793.
37. Uyama, H.; Kobayashi, S. *Chem. Lett.* **1993**, 1149.

38. Uyama, H.; Takeya, K.; Hoshi, N.; Kobayashi, S. *Macromolecules* **1995**, *28*, 7046.
39. Uyama, H.; Kikuchi, H.; Takeya, K.; Hoshi, N.; Kobayashi, S. *Chem. Lett.* **1996**, 107.
40. Uyama, H.; Takeya, K.; Kobayashi, S. *Bull. Chem. Soc. Jpn.* **1995**, *68*, 56.
41. Kobayashi, S.; Shoda, S.; Uyama, H. *Adv. Polym. Sci.* **1995**, *121*, 1.
42. MacDonald, R.T.; Pulapura, S.K.; Svirkin, Y.Y.; Gross, R.A.; Kaplan, D.L.; Akkara, J.; Swift, G.; Wolk S. *Macromolecules* **1995**, *28*, 73.
43. Matsumura, S.; Beppu, H.; Tsukada, K.; Toshima, K. *Biotech. Lett.* **1996**, *18*, 1041.
44. Namekawa, S.; Uyama, H.; Kobayashi, S. *Polym. J.* **1996**, *28*, 730.
45. Svirkin, Y.Y.; Xu, J.; Gross, R.A.; Kaplan, D.L.; Swift, G. *Macromolecules* **1996**, *29*, 4591.
46. Nobes, G.A.R.; Kazlauskas, R.J.; Marchessault, R.H. *Macromolecules* **1996**, *29*, 4829.
47. Matsumura, S.; Beppu, H.; Nakamura, K.; Osanai, S.; Toshima, K. *Chem. Lett.* **1996**, 795.
48. Ito, S.; Naio, S.; Unemoto, T. *J. Jpn. Oil Chem. Soc. (Yukagaku)* **1988**, *37*, 1006.
49. Matsumura, S.; Shimokobe, H. *Chem. Lett.* **1992**, 1859.
50. Matsumura, S.; Kurita, H.; Shimokobe, H. *Biotech. Lett.* **1993**, *15*, 749.

Chapter 5

Monomer and Polymer Synthesis by Lipase-Catalyzed Ring-Opening Reactions

Kirpal S. Bisht[1], Lori A. Henderson[1], Yuri Y. Svirkin[1], Richard A. Gross[1,4],
David L. Kaplan[2], and Graham Swift[3]

[1]Department of Chemistry, University of Massachusetts—Lowell,
Lowell, MA 01854
[2]Biotechnology Center, Department of Chemical Engineering, Tufts University,
4 Colby Street, Medford, MA 02155
[3]Rohm and Haas Company, 727 Norristown Road, Spring House, PA 19477

Chemo-enzymatic routes to optically active [R]-poly(α-methyl-β-propiolactone) (PMPL) and [R]-poly(β-methyl-β-propiolactone) (P3HB) from racemic α-methyl-and β-methyl-β-propiolactone (MPL and BL, respectively) were demonstrated. The [R]-enriched monomers were prepared by resolution of lactones using lipases in organic media. Subsequently, these monomers were polymerized by chemical methods. Highly crystalline (approximately 73%) 93%-[R]-PMPL had T_m and ΔH_f values of 131 °C and 22.0 cal/g, respectively. The lipase PS-30 (from *Pseudomonas fluorescens*) was used to catalyze the stereoelective ring-opening polymerization of racemic MPL which gave 75%-[S] enriched PMPL. Based on analysis of chain stereosequence distribution by ^{13}C-NMR there was excellent agreement between experimental results and those calculated by the enantiomorphic-site control model. This is consistent with that the enzyme catalyst controls the stereochemistry of chain propagation. Thus, complementary approaches towards the synthesis of [R]- and [S]-enriched PMPL were devised. Mechanistic investigations of ε-caprolactone (ε-CL) polymerization using porcine pancreatic lipase (PPL) as the catalyst were carried out. For polymerizations with a 15 to 1 monomer/butylamine ratio and low water content, rapid initiation and slow propagation were observed. The presence of water concentrations in polymerization reactions above that which is strongly enzyme bound was an important factor which limited the molecular weight of PCL chains. In addition, the living or immortal nature of the polymerizations was assessed from plots of $\log\{[M]_0/[M]_t\}$ versus time

[4]Corresponding author.
90

and M_n versus conversion. Given the ambiguity of the results with respect to chain transfer, it was concluded that the PPL-catalyzed ring-opening of ε-CL was best described as a 'controlled' polymerization. An expression for the rate of propagation was derived from the experimental data which is consistent with that derived from the proposed enzyme-catalyzed polymerization mechanism. The absence of termination in conjunction with the relationship between molecular weight and the total concentration of multiple initiators suggests that ε-CL polymerization by PPL catalysis shares many features of immortal polymerizations.

Pioneering work by Klibanov (*1,2*) and others (*3-5*) carried out on small molecules has quite elegantly demonstrated that suspensions of commercial enzymes in anhydrous organic solvents are powerful catalysts for a range of enantio- and regioselective transformations. The use of enzymes in organic media has shown promising substrate conversion efficiency, high enantioselectivity, and offers advantages relative to reactions in aqueous media such as catalyst recyclability, increased enzyme thermal stability, solubility of a wide range of substrate types in the reaction media and no requirement for pH adjustment as the reaction proceeds (*6,7*). Lipases in organic media have been used successfully for the preparation of enantioenriched γ-butyrolactones, ω-lactones and δ-lactones by lactonization of racemic γ-hydroxy esters, ω-hydroxy esters and δ-hydroxy esters, respectively (*8-10*). From review of this literature it was clear that there are many opportunities where enzyme technology can be exploited to solve problems not only in monomer preparations but also in enantio- and regioselective polymer synthesis.

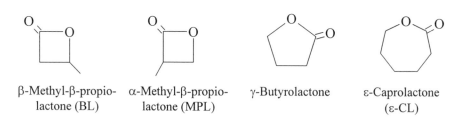

β-Methyl-β-propio- α-Methyl-β-propio- γ-Butyrolactone ε-Caprolactone
lactone (BL)   lactone (MPL)   (ε-CL)

Early work by Klibanov and co-workers(*11*) demonstrated the potential of enzyme catalysts for use in polymerization reactions. These workers prepared chiral oligoesters by lipase-catalyzed reactions between a racemic diester and an achiral diol, or a racemic diol and an achiral diester (*11*). This strategy for enantioselective polyester synthesis was further developed by Wallace and Morrow (*12*). In their research, enantioenriched polyesters were synthesized by the polymerization of bis(2,2,2-trichloroethyl) trans-3,4-epoxyadipate with 1,4-butanediol using porcine pancreatic lipase (PPL) as catalyst (M_w = 7900 g/mol, >95% optical purity). Gutman and co-workers (*13*) also reported the enzymatic synthesis of chiral oligoesters (ee ~34%) from the prochiral substrate dimethyl β-hydroxyglutarate.

In our work, lipases were also selected as catalysts for enantio-and regioselective transformations. Goals of our work included the use of lipases to establish important new routes to chiral polyesters as well as other novel polymeric structures. Critical considerations in the choice of the systems studied were as follows i) to decrease the number of synthetic steps required for the preparation of monomers in high enantiopurity, ii) develop methods that could be applied to a number of related structures and iii) to look at lipases as a relatively new class of polymerization catalysts which require fundamental study to understand their potential as polymerization catalysts.

PHB   **PMPL**

Poly(β-hydroxyalkanoates) (PHAs) represent an important class of natural polyesters that are produced by bacteria. These polymers are enantiopure ([R]-configuration repeat units) and, therefore, isotactic. A number of reviews have appeared in the literature that describe the range of structures obtained, bacterial production systems, fermentation methods, polymer properties and applications (*14-17*). Thus far, chemical routes to enantiopure β-substituted-β-lactone monomers for PHA synthesis have either involved tedious multistep synthetic routes (*18-23*), the use of diketene which presents difficulties in handling and toxicity (*24,25*), or lack flexibility for the preparation of enantiopure monomer structural analogs. Several other optically active poly(α-substituted-β-propiolactone) have been prepared via ring-opening polymerization of enantioenriched lactone monomers. Examples include poly(α-phenyl-β-propiolactone) (*26*), poly(α-ethyl-α-phenyl-β-propiolactone) (*27,28*), and poly(α-methyl-α-ethyl-β-propiolactone) (*29*). Similar to the discussion above for β-substituted-β-lactone monomers, the chemical methods used to prepare α-substituted-β-lactone monomers have involved similar difficulties (*30*). Enzymes could overcome these obstacles to provide: i) a more efficient means of monomer preparation, ii) enhance the enantiomeric enrichment in the final polymer.

An interesting route to enzyme-catalyzed polymer forming reactions is that of ring-opening chemistry. Inherent advantages of this approach are that upon monomer ring-opening, a leaving group is not produced that would lead to reactions of polymer degradation, the ring-strain of the system can be altered to modulate polymerization kinetics, and chiral lactones can be polymerized by enantioselective mechanisms. It is now known that lipases will catalyze the ring-opening of several lactone monomers. Early work by Knani *et al.* (*31*) showed that PPL could be used to catalyzed ε-caprolactone (ε-CL) ring-opening polymerization (624 h, 40 °C). High monomer conversion was found and oligomeric poly(ε-caprolactone), PCL (M_n = 1900 g/mol) was formed. Kobayashi and co-workers (*32*) have reported the polymerization of γ-valerolactone and ε-CL. Of the lipases studied, that from *P. fluorescens* (PS-30) gave the highest percent ε-CL conversion (92%) and PCL molecular weight (M_n = 7700 g/mol; degree of polymerization, DP = 68). Polymerization of β-butyrolactone (BL), γ-butyro-lactone, β-propiolactone and ε-CL by PPL and PS-30 lipases was reported to result in low molecular weight polymers (*33*). Recent developments by Kobayashi and co-workers in synthesizing polyesters from the macrolactones 11-undecanolide and 15-pentadecanolide (*34*) demonstrate that the rates of reactions and molecular weights can be greatly enhanced by the proper selection of enzyme-substrate pairs. In addition to work directed at monomer-enzyme specificity, work is currently underway in our laboratories where the manipulation of reaction parameters is used to directly accelerate chain propagation. By this strategy we expect to achieve enhanced enzyme-catalyzed polymerization rates and high molecular weight products. Of course, such research requires study of the polymerization mechanism. As a model system for these investigations we focused on ring-opening polymerizations using the monomer-catalyst pair ε-CL/PPL.

In summary, this chapter reviews recent work carried out in our laboratory which was directed at novel strategies for using enzyme technology to prepare chiral polyesters and for 'controlled' ring-opening polymerizations. It is divided into the following three subtopics: (1) lipase resolution of β-lactones to obtain monomers in high enantiopurity for subsequent chemical polymerization, (2) the use of enzymes as catalysts for stereoelective ring-opening polymerizations and (3) investigation of the propagation kinetics and mechanism of ε-CL enzyme-catalyzed polymerization.

1.0 Chemoenzymatic Routes to Polyesters: Lipase Resolution of β-Lactones and Their Subsequent Chemical Polymerization

1.1 Lipase-catalyzed Resolution of Lactones. The feasibility of using enzyme-catalysis in organic media for the resolution of substituted β-propiolactones enantiomers was investigated in our laboratory. The resolved lactones could then be polymerized via chemical methods to form substituted poly-β-esters having the desired stereochemical purity and configuration. For this study, we chose racemic BL and α-methyl-β-propiolactone (MPL) and evaluated various lipases including porcine pancreatic lipase (PPL), the lipase from *Candida cylindracea* (CCL) and the *Pseudomonas fluorescens* lipase (PS-30). The effect of the lipase-organic solvent pairs used on the rate of BL reactivity and enantioselectivity was determined. By using enantioenriched lactone prepared by enzymatic resolution, P3HB having 90%-[R] and PMPL having 93%-[R] repeat units were synthesized (see below).

The effect of the organic solvent used on the PPL-catalyzed conversion of substituted-β-propiolactones (BL and MPL) to their corresponding methyl and benzyl esters was studied. For reactions carried out in different anhydrous organic solvents, lactone conversion occurred at rates that were in the following order (faster to slower): n-hexane>diethyl ether>1,4-dioxane ≅ ethyl acetate > benzene. Figure 1 shows monomer conversion as function of time for the lipase-catalyzed methanolysis carried out in diethylether. When these reactions were conducted without enzyme addition, no ester formation was observed (by FTIR) within a 48 h time period.

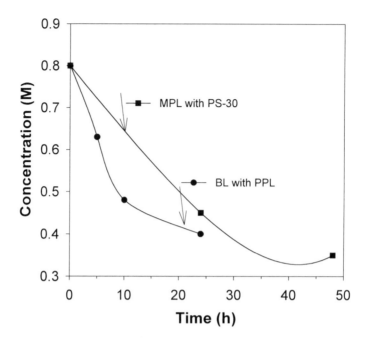

Figure 1. Lipase catalyzed methanolysis in diethylether at 35 °C: lactone consumption as a function of time (MeOH:lactone 2:1).

The enantiomeric ratio E is a measure of the enantioselectivity of an enzymatic transformation. Values of E were determined by equation 1 below based on relationships developed by Chen et al. (*35,36*).

$$E = \frac{V_{maxA} / K_A}{V_{maxB} / K_B} = \frac{\ln[(1-c)(1-ee)]}{\ln[(1-c)(1+ee)]} \tag{1}$$

The absolute configuration of the enantioenriched lactone stereoisomer (slower reacting enantiomer) for BL and MPL in all of the solvents investigated was [R] based on the positive optical rotation (*18*) and the relatively higher intensity of upfield doublets corresponding to the [R]-antipode in ^{1}H NMR experiments with chiral shift reagent Eu-(+)-[hfc]₃ (*20,37,38*). Using equation 1 and a curve-fitting computer program, theoretical curves were generated that fit well with the experimental ee versus conversion results, suggesting a mechanism that agrees with the assumptions used for the derivation of equation 1 (*35,36*). The E values from the curve-fit of the experimental results for BL and MPL resolution reactions carried out in diethyl ether, benzene, ethyl acetate, 1,4-dioxane and hexane are tabulated in Table I.

Table I. Enantiomeric ratio E values for lipase-catalyzed methanolysis
reactions of BL and MPL.[a]

enzyme	lactone	diethyl ether	benzene	ethyl acetate	dioxane	hexane
PPL	BL	20.4	14.8	8.94	8.15	5.56
	MPL	3.93	2.98	2.99	2.52	1.12
PS-30	BL	5.2	3.8	3.8	2.7	7.0
	MPL	8.1	18	5.5	6.1	n/a

[a] 35° C, [MeOH]/[lactone] =2(molar ratio), lactone/solvent = 0.92 mmol/mL, reaction time was adjusted so that conversion was between 35 and 50%.

Since resolutions in hexane were accompanied by oligomerization and recalling from above that diethyl ether gave relatively high rates of lactone conversion, we concluded from the current data base that diethyl ether was the preferred organic medium to achieve high enantioselectivity as well as conversion rates. It is interesting to note that the E values for resolutions of MPL using PPL were much smaller, regardless of the organic media used (*37*), than that obtained with PS-30. In contrast, for MPL resolution, PS-30 was preferred relative to PPL due to the higher enantioselectivity achieved (*37*). The E value for resolutions with benzyl alcohol was relatively lower than that obtained when methanol was used (8.0±1.0 as compared to 20.4±2.4 for BL resolution).

Resolutions carried out with the lipase CC gave relatively poorer results. For example, CC showed no apparent catalytic activity for MPL or BL methanolysis in diethyl ether at 35 °C for 48 h. Resolutions with PPL and PS-30 lipases were used to prepare 90%-[R]-BL and 93%-[R]-MPL, respectively (*37,38*).

1.2 Polymerization of Enantioenriched Lactone Monomers. The 90%-[R]-enriched BL was used to prepare enantioenriched P3HB following a literature method (*20*). The $Zn(C_2H_5)_2/H_2O$ (1/0.6) catalyst system used was believed to result in ideal random stereocopolymers (*39*) with retention of configuration and no apparent racemization (*20*). Thus, the product formed presumably has a stereochemical composition equivalent to the monomer feed. The P3HB (90%-[R]-P3HB, M_n = 41,300 g/mol and M_w/M_n = 1.57 by GPC) had a T_m and ΔH_f of 140 °C and 15.4 cal/g, respectively. In a similar fashion, mixtures of 93%-[R]-MPL and racemic-MPL were used to prepare PMPL of variable enantiopurities (Table II) by anionic polymerization using CH_3COOK/dibenzo-18-crown-6 as initiator. Thus, PMPL having predominantly [R]-repeat unit stereochemical configurations was formed and the polymer stereochemical purity was assumed to be equivalent to that of the monomer feed.

WAXS data collected for 90%-[R]-PMPL which was crystallized by annealing from the melt at 100 °C showed well resolved crystalline reflections superimposed on a low intensity amorphous halo. This indicates high levels of crystallinity and uniformly organized crystals. The degree of crystallinity, χ_c, was estimated to be 73%, similar in value to natural P3HB (60 to 74%) (*40*).

Table II. Preparation[a] of [R]-enriched PMPL stereoisomers and product characterization by gel permeation chromatography (GPC) and differential scanning calorimetry (DSC).

Product	[M]/[I]	Yield (%)	M_n^b (g/mol)	M_w/M_n^b	T_g^c (^{o}C)	T_m^c (^{o}C)	ΔH_f^c (cal/g)
50%-[R]-PMPL	500	88	11,500	1.12	-30	-	-
58%-[R]-PMPL	550	80	14,700	1.22	-27	42	3.8
71%-[R]-PMPL	1080	91	16,300	1.24	-25	65	10.0
74%-[R]-PMPL	450	90	10,600	1.25	-26	77	14.7
90%-[R]-PMPL	500	86	12,800	1.36	-25	130	20.7
93%-[R]-PMPL	750	95	9,300	1.59	-24	131	22.0

[a] The [R]-enriched MPL was resolved by the lipase PS-30; CH_3COOK/dibenzo-18-crown-6 (1:2) was used as the initiator for [R]-enriched MPL ring-opening polymerization (60 ^{o}C, 4 days, in bulk). [b] Molecular weight measurements were by GPC relative to polystyrene standards. [c] Measurements of thermal transitions were by DSC.

Table III. Chain sequences of stereocopolymers 50%-, 74%- and 93%-[R]-PMPL determined by ^{13}C NMR.

product	diad[a,c]		triad[b,c]			$E^{d,e}$
	(i)	(s)	(ii)	(ss)	(is+si)	
50%-[R]-PMPL	0.49	0.51	0.24	0.26	0.50	-
	(0.50)	(0.50)	(0.25)	(0.25)	(0.50)	
74%-[R]-PMPL	0.71	0.29	0.54	0.15	0.31	0.97
	(0.62)	(0.38)	(0.42)	(0.19)	(0.39)	
			[0.56]	[0.15]	[0.29]	
93%-[R]-PMPL	1.00	0.00	1.00	0.00	0.00	-
	(0.87)	(0.13)	(0.80)	(0.07)	(0.15)	
			[1.00]	[0.00]	[0.00]	

[a] Diad fractions i (isotactic) and s (syndiotactic) were calculated from the relative areas of the carbonyl ^{13}C NMR signals. [b] Triad fractions ii (isotactic), ss (syndiotactic) and $is+si$ (heterotactic) were calculated from the relative areas of the methine ^{13}C NMR signals. [c] Values in parentheses were calculated from the enantiopurities of the monomer feeds using Bernoulli equations (41). Values in brackets were calculated using the enantiomorphic-site model equations (42-44). [d] Enantiomorphic-site model triad test $E=2(ss)/(is+si)=1$ for a polymer perfectly described by the enantiomorphic-site model. [e] E value for 93%-[R]-PMPL can not be calculated due to the absence of the s and $is+si$ peaks.

Analysis of PMPL stereoisomer diad and triad sequence distributions were carried out by ^{13}C NMR. Comparison of experimental and calculated (by the Bernoulli model) sequence fractions for the 50%-[R]-PMPL sample were in excellent agreement. In contrast, for the 74 and 93%-[R]-PMPL samples, comparison of the experimental diad and triad fractions with those calculated by the Bernoulli model showed that chain propagation no longer proceeds by ideal random propagation statistics. In fact, there appears to be a tendency toward isospecificity. The triad data for these latter two samples were also evaluated by the enantiomorphic-site model (see Table III, legends c and d). Interestingly, using this model, it was found that there was excellent agreement between calculated and experimental triad fractions. These observations for polymerizations of 74%- and 93%-[R]-MPL were discussed in greater detail elsewhere (*38*).

2.0 Enzyme-Catalyzed Stereoelective Ring-Opening Polymerization

From our work with the resolution of MPL in hexane by PS-30, even though the molar ratio of methanol to MPL was high (2 to 1), it was observed that substantial quantities of oligomer was formed. In other words, MPL ring-opening products appeared to be favorable nucleophilies for MPL ring-opening resolution. Therefore, work was undertaken to investigate whether, instead of carrying out a two-step process of monomer resolution and subsequent polymerization, it might be possible to directly obtain enantioenriched PMPL via a stereoelective polymerization catalyzed by PS-30. This took on additional interest since, to our knowledge, an enzyme catalyzed stereoelective ring-opening polymerization had not been documented. Previous attempts at stereoelective polymerization of racemic lactones have typically involved the use of chiral organometallic catalyst systems and have led to only modest enantiomeric enrichment of the final products (*45,46*).

Polymerizations of MPL catalyzed by PS-30 were carried out in heptane as well as toluene and dioxane, and the effects of experimental parameters on the kinetics and enantioselectivity of PS-30 catalyzed racemic MPL polymerizations were studied. Control reactions maintained under the same conditions as above but without the addition of enzyme did not show noticeable conversion of MPL. The optically active polymers formed were characterized to determine product molecular weight averages by GPC, thermal properties by DSC, repeat unit sequence distribution by ^{13}C NMR and end group structure by ^{1}H and ^{13}C NMR.

Values of monomer conversion were obtained by the relative intensities of PMPL methyl (C$\underline{H}_3$-CH, doublet at 1.15 ppm) and MPL methyl (C$\underline{H}_3$-CH, doublet at 1.35 ppm) signals (*47*). Results of replicate experiments carried out in toluene-d_8 and dioxane-d_8 showed that monomer conversion to PMPL proceeds at much faster rates in the former solvent.

End group structural analysis of PMPL chains formed by enzyme-catalyzed ring-opening polymerization was carried out by ^{1}H and ^{13}C NMR. Spectra were recorded before and after diazomethane derivatization for study of terminal carboxyl chain ends. It was concluded that the chains formed have alcoholic and acidic chain end structures. This is consistent with previous work on lipase catalyzed CL ring-opening polymerizations (*32,48*).

Table IV. Effect of solvent on monomer conversion and PMPL molecular weight.

Product[a]	Solvent	Conversion (%)[b]	M_n^c, (g/mol)	M_n^d, (g/mol)	M_w/M_n^e
T144-1	Toluene	71	2600	2000	1.7
T72-2	Toluene	41	2900	2200	1.8
T96-3	Toluene	47	2700	2200	1.8
D744-1	Dioxane	37	600	500	1.3
D96-1	Dioxane	28	400	400	1.1
H144-1	Hexane	63	2700	1900	1.8

[a] Letter in abbreviations corresponds to the solvent used (T, toluene; D, dioxane; H, heptane) followed by polymerization time in hours and the experimental run. [b,c] Determined by ^{1}H NMR measurements (*47*). [d,e] Determined by GPC analysis relative to polystyrene standards.

Number average (M_n) molecular weight values and polydispersities (M_w/M_n) of the synthesized polymers are shown in Table IV. For purposes of consistency, the values for M_n reported in the text below are from ^{1}H NMR measurements. PMPL products obtained using toluene as the organic medium (T144-1, 71% conversion; T72-2, 41% conversion; T96-3, 47% conversion) had relatively similar M_n (2600, 2900 and 2700 g/mol, respectively) and M_w/M_n (1.7, 1.8 and 1.8, respectively) values. That M_n did not increase at higher conversion (71%) may be due to enzyme catalyzed chain hydrolysis reactions. PMPL synthesized in heptane (H144-1) had similar M_n and M_w/M_n values relative to those prepared in toluene (Table IV). In contrast, the M_n values of products prepared in dioxane (D744-1 and D96-1, % conversions 37 and 28, respectively) were relatively low (600 and 400 g/mol, respectively). Once again, this was rationalized as resulting from enzyme-catalyzed chain hydrolysis which would be increasingly important for the relatively slower propogation rates of polymerizations conducted in dioxane as opposed to toluene or heptane.

The enantioselectivity of PS-30 catalyzed MPL polymerizations was determined based on knowledge of monomer conversion and the stereochemical purity of unreacted MPL. At the termination of polymerizations (~50% conversion), remaining MPL was purified by column chromatography (*47*). The optical purity of unreacted MPL was determined by ^{1}H NMR using (+)-Eu(hfc)₃ as a chiral shift reagent. Complexation of MPL stereochemical mixtures with (+)-Eu(hfc)₃ causes unequal downfield shifts of doublets corresponding to the methyl protons in [S]- and [R]-MPL isomers (*38*). Relative integration values of the [R]- and [S]-MPL doublets were used to calculate residual monomer enantiomeric composition ([S] mol fraction = [S]/{[R]+[S]}) and enantiomeric excess (ee, [R]-[S]/[R]+[S]) values. From information on unreacted monomer, enantiomeric composition and conversion, values for PMPL enantiomeric composition and ee were determined and are compiled in

Table V (see footnotes for equations used). Thus, considering both polymerization reaction rates and enantioselectivity, toluene was found to be the preferred solvent for preparing PMPL enriched in the [S] enantiomer.

Table V. Effect of Solvent and Conversion on PMPL Optical Purity.

Product[a]	Conv. (%)	MPL $[S/(R+S)]_m$	E	$(ee)_m$, MPL (unreacted)[b]	PMPL $[S/(R+S)]_p$[c]	$(ee)_p$, PMPL[b]	$[\alpha_0]_D^{25}$, (g/dL) PMPL
T144-1	71	0.08	4.2	0.84	0.67	0.34	+12.2
T96-3	47	0.28	4.7	0.45	0.75	0.50	+19.0
D744-1	37	0.42	2.0	0.16	0.63	0.26	+7.2
H144-1	63	0.41	0.9	0.19	0.62	0.24	+4.3

[a] See Table IV, footnote *a*. [b] Enantiomeric excess was determined as $(ee)_m=[(R-S)/(R+S)]_m$ for unreacted MPL and $(ee)_p=[(S-R)/(R+S)]_p$ for PMPL. [c] PMPL composition was calculated by equations: $S_p=0.5-\{(1-C)\,[S/(S+R)]_m\}$ and $R_p=0.5-\{(1-C)\,[R/(S+R)]_m\}$, where C is conversion and $[S/(R+S)]_m$ and $[R/(S+R)]_m$ are the fractions of [S] and [R] enantiomers in unreacted MPL.

Figure 2 shows experimental conversion versus ee (shown as inverted triangles) for MPL resolution via PS-30 catalyzed polymerization carried out in toluene-d_8.

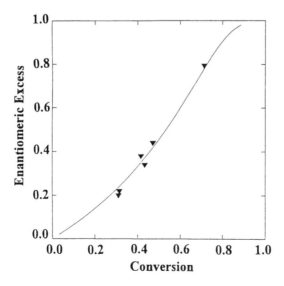

Figure 2. Experimental ee vs %-conversion (denoted as inverted triangles) for a polymerization carried out in toluene-d_8 along with a theoretical curve generated using equation 1.

Equation 1 was used to generate the corresponding theoretical curve shown in Figure 2. Since the theoretical curve is in close agreement with experimental data points, it is reasonable to assume that this polymerization occurs by a mechanism that agrees with the assumptions used for the derivation of equation 1 (*35*).

Table VI. Diad and Triad Stereosequence Distributions and Statistical Analyses for PMPL Samples.

| Product | S/(S+R) | Diad fraction[a] | | Triad fraction[d] | | | B[e] | E[f] |
		(i)	(s)	(ii)	(ss)	(is+si)		
T144-1	0.67	0.67[b]	0.33	0.40	0.23	0.37		
		0.64[c]	0.36	(0.40)[g]	(0.12)	(0.48)	2.68	
				[0.46][h]	[0.18]	[0.36]		1.24
T96-3	0.75	0.74	0.26	0.50	0.16	0.34		
		0.69	0.31	(0.48)	(0.10)	(0.42)	2.77	
				[0.53]	[0.16]	[0.31]		0.94

[a] Given as the fraction of (i) isotactic and (s) syndiotactic diads as determined by ^{13}C NMR, [b] from carbonyl carbon resonance, [c] from methyl carbon resonance. [d] Given as a fraction of (ii), (ss) and (is)+(si) triads as determined by ^{13}C NMR from carbon methine resonance. [e] The Bernoulli model triad test, where $B=4(ii)(ss)/(is+si)^2$, such that B=1 for ideal random propagation. [f] The enantiomorphic-site model triad test, where $E=2(ss)/(is+si)$, such that E=1 for perfect catalyst control. [g] The numbers in parentheses are the calculated triad stereosequences from the experimentally determined diad stereosequence values by using the following Bernoullian equations: $(ii)=(i)^2$, $(ss)=(s)^2$, $(is+si)=2(s)(i)$. [h] The numbers in brackets are the calculated triad stereosequences from the experimentally determined diad stereosequence values by using corresponding relations between triad/diad sequences for enantiomorphic site model (*43,49*): (ii)=1-3(s)/2; (is+si)=(s); 2(ss)=(s).

Experimental results obtained for Products T144-1 and T96-3 diad and triad fractions are shown in Table VI. The triad distribution calculated from experimental diad intensities using the Bernoulli model (*49*) deviated significantly from experimental values. Also, the Bernoulli model triad test where $B=4(ii)(ss)/(si+is)^2 =1$ led to calculated B values for T144-1 and T96-3 of 2.68 and 2.77, respectively, which are significantly higher than 1 for ideal random chain propagation (*41,50,51*). Interestingly, calculation of triad stereosequences from experimental diad fractions using the enantiomorphic site model (see Table VI legend h and refs *43, 44, 49*) resulted in better agreement between experimental and calculated triad fractions compared to the Bernoulli model. Furthermore, calculation of the test parameter E = (2ss)/(is+si) based on the enantiomorphic site model (*52*) gave values for T144-1 and T96-3 of 1.24 and 0.94 which is in good agreement with the theoretically predicted value of 1.0. Indeed, isospecific polymerization controlled by catalyst chirality usually follows the enantiomorphic site control model (*43*). This model assumes that catalyst structure and not the structure of polymer chain end controls the stereochemical configuration of the corresponding polymer. Also, this model assumes that the catalyst structure remains unchanged prior to each new monomer propagation step. Agreement with the enantiomorphic site control model is consistent with our expectations for PS-30 catalyzed lactone polymerization where the enzyme reacts more rapidly with [S]- relative to [R]-MPL, forming an enzyme activated monomer (EAM) complex which then reacts with a growing PMPL chain end. The stereochemistry of the added monomer is controlled by PS-30 during formation of the PS-30-MPL EAM complex (Scheme 1).

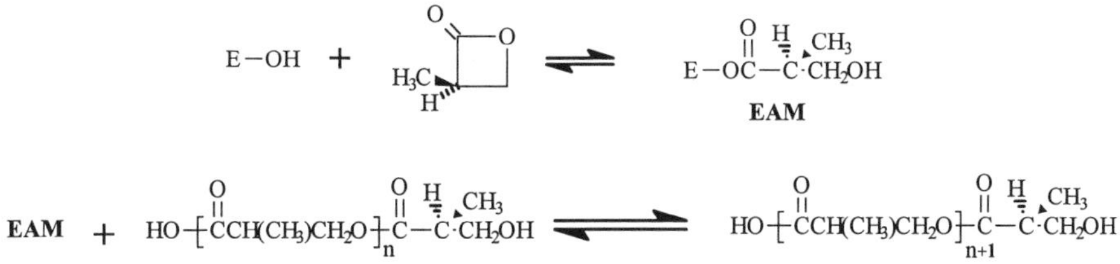

$$E-OH \; + \; \underset{H}{\overset{O}{H_3C}}\underset{}{\overset{}{\bigcirc}}\!\!-O \;\; \rightleftharpoons \;\; E-O\overset{O}{\overset{\|}{C}}-\overset{H}{\overset{\text{\tiny=}}{C}}\cdot CH_2OH \quad \underset{\textbf{EAM}}{\,}$$

$$\textbf{EAM} \; + \; HO \!-\!\!\left[\overset{O}{\overset{\|}{C}}CH(CH_3)CH_2O\right]_n\!\!\overset{O}{\overset{\|}{C}}-\overset{H}{\overset{\text{\tiny=}}{C}}\cdot CH_2OH \;\; \rightleftharpoons \;\; HO \!-\!\!\left[\overset{O}{\overset{\|}{C}}CH(CH_3)CH_2O\right]_{n+1}\!\!\overset{O}{\overset{\|}{C}}-\overset{H}{\overset{\text{\tiny=}}{C}}\cdot CH_2OH$$

Scheme 1

Furthermore, since no other initiator species was added to the polymerizations and PMPL chains were formed having hydroxyl and carboxylic chain ends (see above), it is presumed that water functioned as the nucleophile for chain initiation.

3.0 Mechanistic Investigations of Lipase-Catalyzed ε-Caprolactone (ε-CL) Ring-Opening Polymerization

In general, enzyme-catalyzed polymerizations require long reaction times for complete monomer conversion and have resulted in low number average molecular weight (M_n) polyesters (*31-33*). In addition, the polymerizations normally proceeded with low rates of propagation. Thus, we have initiated a program to better understand the mechanism of enzyme-catalyzed polymerization. Such information is expected to lead to enhanced propagation kinetics and a decrease in reactions that limit product molecular weight.

Two investigations were performed exploring the ring-opening of ε-CL using PPL as the catalyst. Initially, systematic experiments were designed to determine the effects of reaction parameters such as the enzyme water content, butanol chain initiation and the organic media used on monomer conversion, product molecular weight and end-group structure (*48*). The affects of supplementing reactions with butanol demonstrated that nucleophile addition to polymerizations could be used to 'tailor' chain end structure. Also, the potential to achieve 'controlled' enzyme-catalyzed ring-opening polymerizations (i.e., molecular weight controlled by monomer/initiator stoichemetry) was realized. Based on these findings, we used the PPL-catalyzed ring-opening of ε-CL as a model system for a comprehensive investigation of polymerization kinetics (*53*). Three nucleophiles (butanol, water and butylamine) were used to study their affects on chain initiation and propagation. The rate expression for chain propagation was determined from assessments of living and immortal characteristics. The results of this work were used to discuss mechanistic features of lipase-catalyzed ring-opening polymerizations. The experimental protocols for determinations of product structure, polymer molecular weight, total chain number and the mol fraction of end groups were described elsewhere (48,53).

3.1 Effects of nucleophile and water concentration on PPL-catalyzed ε-CL ring-opening polymerizations. Polymerizations of ε-CL catalyzed by PPL in heptane at 65 °C were supplemented with butanol to observe the effects on initiation and PCL chain-end structure. Products were analyzed by ^{1}H-NMR to determine the chain end-group structure and M_n. For reactions with a ε-CL to butanol ratio of 14 to 1, the resulting PCL products had a M_n of 1600 g/mol and were found to have butyl ester or carboxylic acid and hydroxyl chain ends. The presence of carboxylic acid chain ends was concluded by derivatizing with diazomethane to produce methyl ester end groups. Thus, water and butanol acted as competitive nucleophiles for chain initiation. This was also confirmed by conducting reactions at various butanol/water ratios and then determining the corresponding mol fractions of butyl ester and carboxylic acid end groups.

3.2 Effects of reaction water concentration on PCL molecular weight. In principle, based on the above and the use of purified PPL, if the quantity of water in the polymerization reaction is decreased to a value where the remaining water is tightly bound to the enzyme and has little to no reactivity, it should be possible to obtain products where the chains are initiated by only the specific initiator added. This would provide a route for 'tailoring' of chain end structure and, as a consequence, preparing a family of macromers. Also, it is likely that high molecular weight PCL was not obtained in this work (48) due to the presence of water in excess of that needed to maintain enzyme activity.

3.3 Kinetics and mechanism of PPL-catalyzed ε-CL polymerization. Propagation kinetics were assessed by conducting reactions to complete monomer conversion and determining: i) monomer conversion as a function of time, ii) values of chain molecular weight, total chain number and flux in the mol fraction of end groups as a function of conversion. Additional studies were carried out at early stages of the reaction (<10% conversion, 2 hours) to gain information on chain initiation.

For polymerizations in heptane at 65°C which contained 0.13 mmoles of water, either butanol (0.33 mmole) or butylamine (0.30 mmoles) was added as a competitive nucleophile for chain initiation. Initially, the molar ratios of ε-CL/water, ε-CL/butanol and ε-CL/butylamine investigated were 35/1, 14/1 and 15/1, respectively. Both conversion and M_n as a function of reaction time were determined. The curves generated for % conversion versus time (Figure 3) were similar for the three polymerizations (53). Thus, regardless of the initiator employed, the rate of monomer consumption was not affected. Also, monomer consumption was not zero order with respect to monomer concentration.

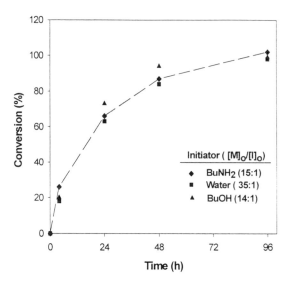

Figure 3. Monomer conversion versus time for three polymerizations carried out in the presence of low water content and with the addition of either butylamine or butanol.

Figure 4 shows a plot of M_n versus %-conversion for the same three systems. The results varied considerably as a function of the initiator system used.

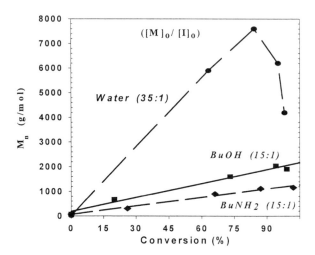

Figure 4. Product molecular weight as a function of %-conversion for polymerizations using butylamine, butanol or only water as nucleophiles.

For all three polymerizations, M_n either decreased substantially or stayed almost unchanged at high conversions (Figure 4). For example, reactions without the addition of butanol or butylamine reached a maximum M_n of 7600 g/mol at 85% conversion and, subsequently, had an M_n of 4200 g/mol at 98% conversion. This is explained by decreased chain propagation rates at high conversion. Thus, at high conversion, competing reactions which decrease M_n such as chain cleavage by hydrolysis become increasingly important relative to propagation. The product polydispersity also increased with %-conversion. Polymerizations carried out without butanol or butylamine had M_w/M_n values as high as 5.0 at 84% conversion.

The relationship between total chain number and %-conversion was shown in Figure 5. For all of these polymerizations, the general trend observed was an increase in the total number of polymer chains, N_p, with conversion.

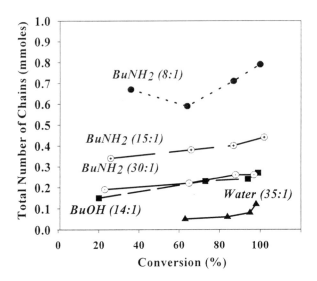

Figure 5. Effects of nucleophile structure and concentration on the total number of chains formed during PPL catalyzed ε-CL polymerizations.

For reactions with butanol and butylamine, this increase in chain number corresponded with a decrease in the mol fraction of butyl ester and butyl amide end groups and an increase in carboxylic acid chain ends (Figure 6). In addition, since the mol fraction of carboxylic acid end groups increases with conversion, we believe that the observed increase in N_p during the polymerization results from reactions with water. Interestingly, based on polymer yield and end group structure, it was found that only 37 and 52% of the butanol was consumed after 4 and 96 h, respectively (i.e., 20 and 100% ε-CL conversion). In contrast, butylamine (ε-CL/BuNH2 ratio 15/1) was completely consumed within 4 h or at 26% monomer conversion. For example, the N_p

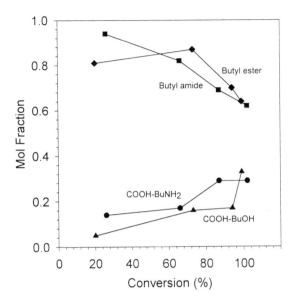

Figure 6. Effects of nucleophile on mol fraction of chain end functional groups during PPL catalyzed ε-CL polymerizations.

at this conversion was equivalent to the amount of butylamine added (0.30 mmoles) and the mol fraction of butyl amide chain ends was 0.94. Thus, butylamine was rapidly consumed (reaction times < 4 h) with a rate of initiation faster than that of propagation. Furthermore, the cumulative results of our work with PPL catalyzed ε-CL polymerization indicates that this is a chain polymerization with slow propagation (*48,53*).

3.4 Products formed at low monomer conversions. Experiments were conducted at early stages of the reaction (0-2 h) to provide information on the relative rates of initiation as a function of the nucleophile added as well as the initial products formed. The polymerizations were carried out in bulk at 65°C and aliquots of the reaction mixture were withdrawn every 0.5 h (*53*). For a 0.5 h reaction time and a ε-CL/butylamine ratio of 15/1, the %-monomer conversion for the control reaction (no enzyme) was negligible (<0.1%) whereas 2.9% monomer was converted with enzyme. At 2 h, monomer conversion for butylamine-initiated reactions without and with enzyme were 1.9 and 7.0%, respectively. The theoretical %-monomer conversion was 7.1% assuming quantitative formation of *N*-butyl-6-hydroxyhexanamide. This value is in excellent agreement with that observed experimentally after 2 h. From the above and careful ^{1}H NMR analysis of reaction products formed within this 2 h reaction period, we concluded the following: i) non-enzyme mediated ring-opening of ε-CL by butylamine occurs but at a rate which is considerably slower than the corresponding enzyme-catalyzed reactions, ii) for reactions with enzyme, butylamine was consumed within 2 h and formed N-butyl-6-hydroxyhexanamide prior to subsequent chain growth.

Similarly, the %-monomer conversion for reactions with butanol was determined for reaction times from 0 to 2 h. After 0.5 and 2 h, 1.7 and 3.1% of ε-CL was converted. Therefore, chain initiation by butylamine was more rapid than butanol. Unfortunately, ^{1}H NMR analysis of the ring-opened products at low conversion did not provide information on the relative distribution of structures (monoadduct versus oligomers).

3.5 Mechanism of butylamine initiation and subsequent propagation for PPL-catalyzed ε-CL polymerization. Based on previous work by others, it is generally assumed that the PPL active site which is responsible for ester hydrolysis has a nucleophilic serine residue (*54,55*). This analysis of PPL suggests that the serine residue acts as the nucleophile for ester hydrolysis and that the serine residue is the catalytically essential region where the enzyme hydrolyzes the ester functionality by attacking the ester carbonyl. We have shown by ^{1}H NMR analyses that PPL-catalyzed ε-CL ring-opening polymerization can take place with initiation by either water, n-butanol or butylamine. From the study of the reactions at low conversion, the following mechanism for PPL catalyzed ε-CL polymerization using butylamine as the initiator was proposed (*48,53*) (Scheme 2):

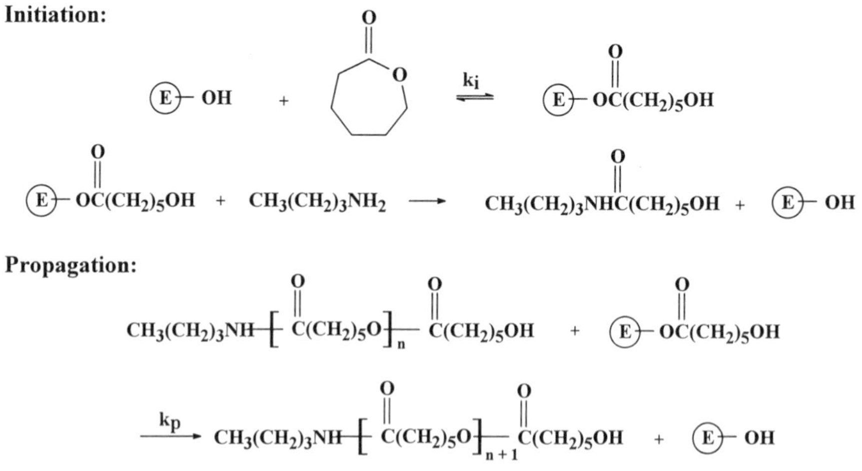

Initiation:

Propagation:

Scheme 2

3.6 Living/immortal characteristics and derivation of a rate expression. Living polymerizations are normally defined as reactions (propagating centers) that do not terminate or undergo chain transfer. In the absence of termination, propagation can be defined by a first order rate law, which upon rearrangement and integration yields a linear expression (*56*). The linearity of M_n as a function of conversion may reflect the lack of chain transfer. Therefore, the living/immortal nature of PPL catalyzed ring-opening polymerizations carried out using butylamine (0.30 mmol, ε-CL/BuNH$_2$ ratio 15/1) in heptane was assessed by constructing plots of $\log\{[M]_0/[M]_t\}$ versus time and M_n versus monomer conversion. For the above, $[M]_0$ is the initial concentration of monomer and $[M]_t$ is the monomer concentration after time t. Since it was demonstrated herein that a considerable amount of monomer was consumed during chain initiation (termed $[M]_i$), the values of $\log\{[M]_0/[M]_t\}$ were corrected and plotted as $\log\{([M]_0-[M]_i)([M]_t-[M]_i)^{-1}\}$ so that monomer consumption during propagation was analyzed. The plot of $\log\{([M]_0-[M]_i)([M]_t-[M]_i)^{-1}\}$ versus time was linear ($r^2 = 0.997$) indicating: *(i)* no termination and *(ii)* that monomer consumption follows a first order rate law.

Plots of M_n versus monomer conversion were also constructed using experimental and calculated M_n values. In this case, calculated M_n values were obtained using equations 2 and 3:

$$M_n = [M]_0/[I]_0 \times C_m \times m_{CL} \qquad (2)$$

$$M_n = [M]_0/[I'] \times C_m \times m_{CL} \qquad (3)$$

where $[I]_0$ is the concentration of butylamine charged, $[I']$ is the total concentration of butylamine and water that reacted (determined by ^{1}H NMR), C_m is the fractional monomer conversion and m_{CL} is the molar mass of ε-CL. The line generated by using equation 2 to calculate M_n deviated significantly from the experimental data. However, there was excellent agreement between the experimental data and the line generated using equation 3 to calculate theoretical M_n values. Knowing that the initiator concentration is best described by $[I']$, the results suggested that chain transfer did not occur. However, for the low M_n values obtained in this work, the affects of chain transfer in plots of M_n versus conversion may not be observed particularly if chain transfer is slow relative to propagation ($k_{tr} \ll k_p$) (*56,57*). Therefore, we conclude that enzyme-catalyzed polymerizations are 'controlled polymerizations'(*58*).

It was also apparent that the rate at which monomer was consumed was independent of the type and concentration of the nucleophile. This implies that the propagation rate, R_p, is independent of $[I]_0$. Furthermore, the type and concentration of nucleophile used had no significant effect on R_p even though N_p varied to a large extent (Figure 5). Therefore, the concentration of chains ends, [R~OH], appears to be zero order with respect to the rate. Based on our results for polymerizations carried out with and without PPL, it was concluded that the rate of monomer consumption was a function of the catalyst concentration (*59*). From these results, an experimental rate expression was defined as: $R_p = k_{obs}[Cat]^x[M]^1[R{\sim}OH]^0$. The first order rate law as indicated by the linear relationship supports the mechanism for propagation discussed above (Scheme 2).

To give a theoretical rate equation from the above propagation mechanism: $R_p = k_p[AM][R{\sim}OH]$, where [AM] is the concentration of the acyl-enzyme intermediate formed in the initiation step. Combining the schemes for chain initiation and propagation, and substituting into the rate equation knowing that [R~OH] is zero order and monomer concentration is first order leads to the rate law $R_p = k_p K[Cat]^x[M]^1[R{\sim}OH]^0$. Rearranging this equation, integrating it with respect to time, and plotting $\log\{[M]_0/[M]_t\}$ versus time should yield a linear relationship with a slope equal to $k_p K[Cat]^x$. Since the slope of the log plot was indeed constant, this indicates that the catalyst concentration is invariable during the course of the polymerization. Thus, the experimentally derived rate expression is consistent with that derived from the proposed mechanism for enzyme-catalyzed ε-CL polymerization.

Enzyme-catalyzed ε-CL polymerization was also compared to the immortal polymerizations described by Inoue (*60*). Interestingly, it was found that when the initiator concentration was taken as the total of the nucleophiles (butylamine and water), $M_n = [M]_0/[I'] \times C_m \times m_{CL}$. This implies that N_p equals the sum of the concentrations of butylamine and reacted water. Moreover, if the initiator concentration $[I]_0$ is assumed to be equal to the concentration of butylamine, the predicted total N_p (i.e., 0.30mmoles at 100% conversion) is less than that actually observed (Figure 5). The M_w/M_n values measured for butylamine reactions were in the range of 1.5 to 2.4 which is considerably greater than that reported by Inoue (~1.13). We believe that broadened molecular weight distributions observed in this work result in part from slow initiation by the nucleophile water as well as by enzyme-catalyzed chain hydrolysis. Furthermore, in regards to exchangeability, we believe that the rate at which propagating PCL chains continue to grow is controlled at the active site through a process of rapid exchangeability of chain ends. This is supported by the fact that accumulation of the monoadduct, *N*-butyl-6-hydroxyhexanamide, was not observed by GPC. Therefore, our data indicates that enzyme-catalyzed polymerizations of ε-CL meets many of the requirements introduced by Inoue (*60*) for immortal polymerizations.

Summary

A chemo-enzymatic route to enantioenriched P3HB which involved resolution of racemic BL and MPL using lipase catalysis and subsequent chemical polymerization of predominantly [R]-lactones was demonstrated. Of the lipases investigated, the preferred enzymes for BL and MPL resolutions were PPL and PS-30, respectively. Using diethyl ether as the organic medium for lipase-catalyzed resolution, 90%-[R]-BL and 93%-[R]-MPL were prepared. By increasing the %-[R]-content in the monomer feed to 93%, a highly crystalline (approximately 73%) PMPL was obtained which, from DSC measurements, had T_m and ΔH_f values of 131 °C and 22.0 cal/g, respectively. In a complementary approach, 75%-[S] PMPL was formed by PS-30 catalyzed ring-opening polymerization of racemic MPL where the [S]-MPL enantiomer reacted more rapidly than its [R]-antipode. As was expected for enzyme-catalysis, analysis of the chain stereosequence distribution by ^{13}C-NMR showed that the polymerization fit the enantiomorphic-site control model.

From the study of reaction conditions on the conversion of ε-CL to PCL using PPL as the catalyst, it was concluded that one can tailor the chain end structure by selecting reactive nucleophiles for ring-opening polymerizations. However, competition for chain initiation also occurs with water. Moreover, the presence of water concentrations in polymerization reactions above that which is strongly enzyme bound is believed to be an important factor which limited the formation of PCL chains of significantly higher molecular weight.

Studies using different nucleophiles for chain initiation showed that the relative rate of initiation was faster for butylamine than butanol or water. Investigations at low conversions showed that butylamine was rapidly consumed to form N-butyl-6-hydroxyhexanamide prior to chain growth. Thus, an activated monomer mechanism was proposed for the ring-opening polymerization of ε-CL catalyzed by PPL. Monomer conversion with time data in conjunction with total number of chains with conversion denotes that this is a chain polymerization with initiation being faster than propagation. The linearity of $\log\{[M]_o/[M]_t\}$ versus time and M_n versus monomer conversion plots indicate: *(i)* a lack of termination, *(ii)* monomer consumption followed a first order rate law and *(iii)* the absence of chain transfer. Given the ambiguity of chain transfer, the enzyme catalyzed polymerizations were described as 'controlled' polymerizations. From the kinetic data it was concluded that the rate of monomer conversion, R_p, was not dependent on: i) the type or concentration of initiator employed and ii) the total number of chains. Thus, an experimental rate expression for R_p was defined which is consistent with that derived from the proposed mechanism for polymerization. In addition, the molecular weight dependence on the total concentration of multiple initiators suggests that ε-CL polymerization by PPL catalysis shares many features with that of immortal polymerizations.

Literature Cited

(1) Klibanov, A. M. *Chemtech.,* **1986,** (June), 354.
(2) Klibanov, A. M. *Acc. Chem. Res.* **1990,** 23, 114.
(3) Dordick, J. S. *Enzyme Microb. Technol.* **1989,** 11, 194.
(4) Halling, P. J. *Biotechnol. Adv.* **1987,** 5, 47.
(5) Lilly, M. D. *J. Chem Tech. Biotechnol.* **1982,** 32, 162.
(6) Zaks, A.; Klibanov, A. M. *J. Am. Chem. Soc.* **1984,** *106*, 2687.
(7) Wilson, W. K.; Baca, S. B.; Barber, Y. J.; Scallen, T. J.; Morrow, C. J. *J. Org. Chem.* **1983,** *48*, 3960.
(8) Gutman, A. L.; Zuobi, K.; Bravdo, T. *J. Org. Chem.* **1990,** *55*, 3546.
(9) Gutman, A. L.; Oren, D.; Boltanski, A.; Bravdo, T. *Tetrahedron Lett.* **1987,** *28*, 5367.
(10)Gutman, A. L.; Zuobi, K.; Bravdo, T. *Tetrahedron Lett.* **1987,** *28*, 3861.
(11)Margolin, A. L.; Crenne, J. Y.; Klibanov, A. M. *Tetrahedron Lett.* **1987,** 28, 1607.
(12)Wallace, J. S.; Morrow, C. J. *J. Polym. Sci., Part A: Polym. Chem.* **1989,** 27, 2553.
(13)Gutman, A. L.; Bravdo, T. *J. Org. Chem.* **1989,** 54, 4263.
(14)Gross, R. A. In *Biomedical Polymers: Designed to Degrade Systems*; Shalaby, S.W., ED.; Hanser Publisher: New York, 1994l; pp. 173-188.
(15)Steinbüchel, A. In *Biomaterials: Novel Materials from Biological Sources*; Byrom, D., Ed.; Stockton Press: New York, 1991; pp. 123-213.
(16)Doi, Y. *Microbial Polyesters*; VCH Pub. Inc.: New York, 1990.

(17)Brandl, H.; Gross, R. A.; Lenz, R. W.; Fuller, R. C. In *Advances in Biochemical Engineering/Biotechnology*; Fiechter Ed.; Springer-Verlag: Berlin, Heidelberg, 1990; Vol. 41, pp. 77.

(18)Kemnitzer, J. E.; McCarthy, S. P.; Gross, R. A. *Macromolecules* **1992**, *25*, 5927.

(19)Agostini, D. E.; Lando, J. B.; Shelton, J. R. *J. Polym. Sci., Part A-1* **1971**, *9*, 2789.

(20)Zhang, Y.; Gross, R. A.; Lenz, R. W. *Macromolecules* **1990**, *23*, 3206.

(21)Voyer, R.; Prudhomme, R. E. *J. Polym. Sci., Polym. Chem. Ed.* **1986**, 24, 2773.

(22)Lavallee, C.; Leborgne, A.; Spassky, N.; Prudhomme, R. E. *J. Polym. Sci., Polym. Chem. Ed.* **1987**, 25, 1315.

(23)Voyer, R.; Prudhomme, R. E. *J. Polym. Sci., Polym. Chem. Ed.* **1988**, 26, 117.

(24)Ohta, T.; Miyake, T.; Takaya, H. *J. Chem. Soc., Chem. Commun.* **1992**, 1725.

(25)Wynberg, H.; Staring, E. G. J. *J. Org. Chem.* **1985**, *50*, 1977.

(26)Hmamouchi, M; Prud'homme, R. E. *J. Polym. Sci. Polym. Chem. Ed.* **1991**, *29*, 1281.

(27)Carriere, F. J.; Eisenbach, C. D. *Makromol. Chem.* **1981**, *182*, 325.

(28)D'hondt, C. G.; Lenz, R. W. *J. Polym. Sci. Polym. Chem. Ed.* **1978**, *16*, 261.

(29)Grenier, D.; Prud'homme, R. E. *J. Polym. Sci. Polym. Chem. Ed.* **1981**, *19*, 1781.

(30)Hmamouchi, M; Prud'homme, R. E. *J. Polym. Sci. Polym. Chem. Ed.* **1988**, *26*, 1593.

(31)Knani, D.; Gutman, A. L.; Kohn, D. H. *J. Polym. Sci., Part A: Polym. Chem.* **1993**, *31*, 1221.

(32)Uyama, H., Kobayashi, S. *Chem. Lett.*, **1993**, 1149.

(33)Nobes,G. A. R.; Kazlauskas, R. J.; Marchessault, R. H. *Macromolecules* **1996**, *29*, 4829.

(34)Uyama, H.; Takeya, K.; Kobayashi, S. *Bull. Chem. Soc. Jpn.* **1995**, 68, 56.

(35)Chen, C. S.; Fujimoto, Y.; Girdaukas, G.; Sih, C. J. *J. Am. Chem. Soc.* **1982**, *104*, 7294.

(36)Chen, C. S.; Wu, S. H.; Girdaukas, G.; Sih, C. J. *J. Am. Chem. Soc.* **1987**, *109*, 2812.

(37)Xu, J.; Gross, R. A.; Kaplan, D. L.; Swift, G. *Macromolecules* **1996**,*.29*, 3857.

(38)Xu, J.; Gross, R. A.; Kaplan, D. L.; Swift, G. *Macromolecules* **1996**, *29*, 4582.

(39)Tanahashi, N.; Doi, Y. *Macromolecules* **1991**, *24*, 5732.

(40)Bluhm, T. L.; Hamer, G. K.; Marchessault, R. H.; Fyfe, C. A.; Veregin, R. P. *Macromolecules* **1986**, *19*, 2872.

(41)The diad and triad fractions for the Bernoulli model can be calculated with knowledge of the monomer stereochemical composition (fraction of [R]- and [S]-MPL) using the following relationships: $i=[R]^2+[S]^2$; $s=2[R][S]$; $ii=[R]^3+[S]^3$; $is=si=ss=[R][S]$.

(42)For the enantiomorphic-site model, the triads were calculated using the following equations: $ii=1-3s/2$; $is+si=s$; $2ss=s$, where i and s were obtained experimentally as described in Table III legend a.

(43)Ewen, J. A. *J. Am. Chem. Soc.* **1984**, *106*, 6355.

(44)Hocking, P. J.; Marchessault, R. H. *Macromolecules* **1995**, *28*, 6401.

(45)Spassky, N.; Leborgne, A.; Momtaz, A. *J. Polym. Sci.: Polym. Chem. Ed.* **1980**, *18*, 3089.

(46) Takeichi, T.; Hieda, Y.; Takayama, Y. *Polymer J.* **1988**, *20*, 159.

(47) Svirkin, Y. Y.; Xu, J.; Gross, R. A.; Kaplan, D. L.; Swift, G. *Macromolecules,* **1996**, *29*, 4591.

(48) MacDonald, R. T.; Pulapura, S. K.; Svirkin, Y. Y.; Gross, R. A.; Kaplan, D. L.; Akkara, J.; Swift, G.; Wolk, S. *Macromolecules* **1995**, *28*, 73.

(49) Shelden, R. A.; Fueno, T.; Tsunetsugu, T.; Furukawa, J. *Polymer Letters* **1965**, *3*, 23.

(50) Bovey, F. A. *Polymer Conformation and Configuration*, Academic Press: New York, 1969.

(51) Sepulchre, M. *Makromol. Chem.* **1988**, *189*, 1117.

(52) Inoue, Y.; Itabashi, Y.; Chujo, R.; Doi, Y. *Polymer* **1984**, *25*, 1640.

(53) Henderson, L. A.; Svirkin, Y. Y.; Gross, R. A.; Kaplan, D. L.; Swift, G. *Macromolecules* **1996**, *29*, in print.

(54) *Lypolytic Enzymes*; Brockerhoff, H.; Jensen, R. G. Eds.; Academic Press: New York, 1974; pp. 74.

(55) Jones, J. B. *Aldrichimica Acta* **1993**, *26*, 106.

(56) Faust, R.; Kennedy, J. *J. Macromol. Sci. Chem.* **1990**, *A27*, 649.

(57) Leleu, J.; Bernardo, V.; Polton, A.; Tardi, M.; Sigwalt, P. *Polymer International* **1995**, 37,

(58) Matyjaszewski, K. *Cationic Polymerizations: mechanisms, synthesis and application*, Marcel Dekker Inc.: New York, 1996.

(59) The rate dependence on the concentration of the catalyst was also observed in experiments where the concentration of PPL was decreased from 750 to 375mg for ε-CL polymerization initiated with water (0.12mmoles). The results for % monomer conversion decreased from 100 to 86% after a 96 h reaction time.

(60) Endo, M.; Aida, T.; Inoue, S. *Macromolecules* **1987**, *20*, 2982.

Chapter 6

Solvent–Enzyme–Polymer Interactions in the Molecular-Weight Control of Poly(m-cresol) Synthesized in Nonaqueous Media

Madhu Ayyagari[1], Joseph A. Akkara[1], and David L. Kaplan[2]

[1]Biotechnology Division, U.S. Army Soldier Systems Command, Natick Research, Development, and Engineering Center, Natick, MA 01760—5020
[2]Biotechnology Center, Department of Chemical Engineering, Tufts University, 4 Colby Street, Medford, MA 02155

Poly(m-cresol) was synthesized by horseradish peroxidase-catalyzed reactions in ethanol/buffer mixtures. Polymer molecular weight was controlled by manipulating the solvent composition. The effect of solvent composition on enzyme activity and polymer solubility was studied to understand the factors affecting polymer molecular weight. Enzyme kinetics revealed the effect of solvent on enzyme activity and substrate partitioning between bulk solvent and the enzyme. Thermal and spectroscopic characteristics of the polymer are discussed.

Polyphenols have long been used as engineering materials and specialty polymers. Commercially available phenolic resins such as novolacs and resoles are primarily used as adhesives and laminates. They are prepared in industry by condensing different proportions of phenol and formaldehyde in the presence of an acid or a base catalyst [1]. However, polyphenol production by alternate processes is desirable due to the toxicity of formaldehyde. Inorganic catalysts can be used to polymerize phenol or 2,6-derivatives of phenol to make poly(phenylene oxides) in the presence of molecular oxygen [2]. Likewise, biological catalysts such as polyphenol oxidases can be used to polymerize phenol and its derivatives or aromatic amines in the presence of hydrogen peroxide or molecular oxygen. The advantages offered by enzymatic processes over chemical processes have been reviewed [3-8]. Additionally, polyphenols prepared by enzymatic processes have extensive conjugation in their structure leading to electrical conductivity under doped conditions and nonlinear optical behavior (Figure 1). As a result, these polymers are expected to find wider applications in the fields of electronics and photonics. Although ferric chloride is capable of coupling phenolic units on the aromatic rings leading to conjugated structures, the resultant product is at best dimers or trimers [unpublished results and 9].

  Peroxidase-catalyzed oxidation of phenols in aqueous media has been known for over 50 years [10]. However, a number of substituted phenolic monomers or aromatic amines have very limited solubility in water. Hence, since the advent of nonaqueous enzymology, phenol polymerization in organic solvents by enzyme-mediated reactions has been well studied to produce different types of higher molecular weight polymers [3-8]. Among a number of polyphenol oxidases, the enzyme extracted from the roots of horseradish is widely used to catalyze the reaction because of its

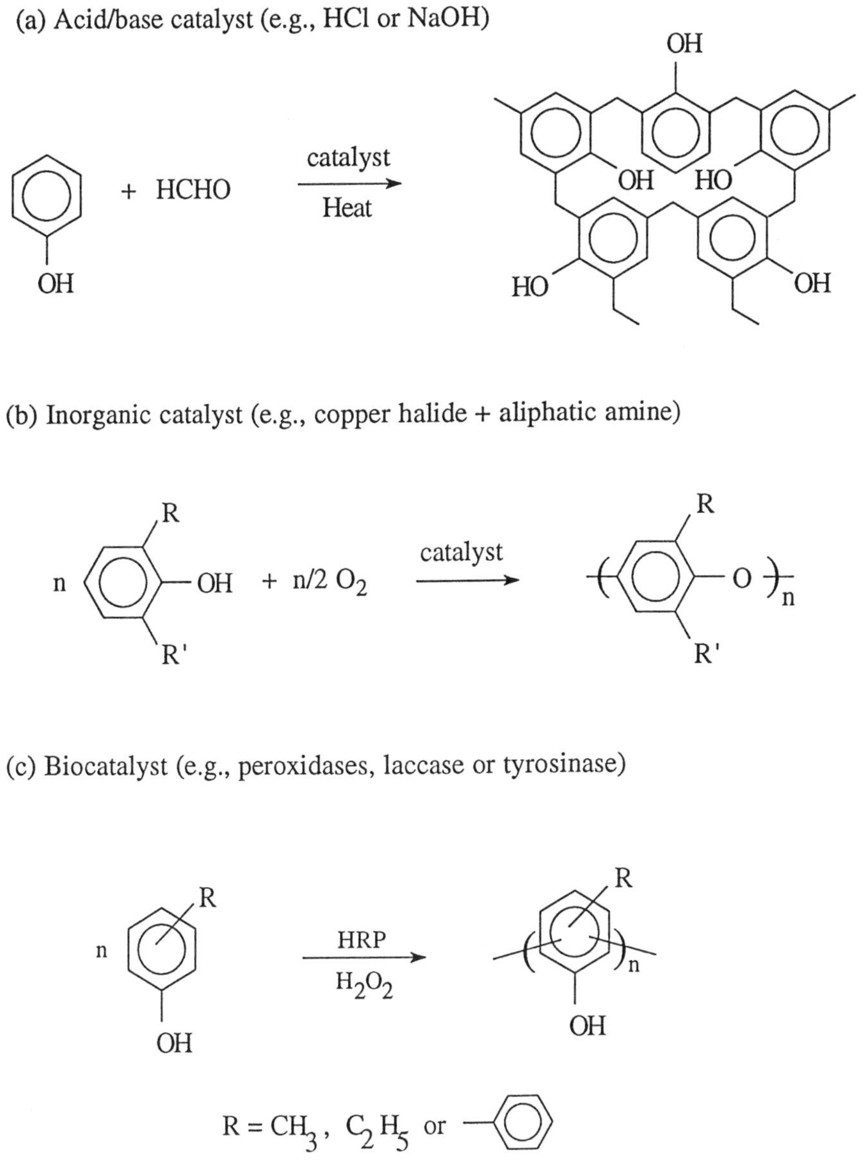
(a) Acid/base catalyst (e.g., HCl or NaOH)
OH
+ HCHO
catalyst
Heat
OH
OH
HO
HO
OH
(b) Inorganic catalyst (e.g., copper halide + aliphatic amine)
n
R
OH
R'
+ n/2 O2
catalyst
R
O
n
R'
(c) Biocatalyst (e.g., peroxidases, laccase or tyrosinase)
n
R
OH
HRP
H2O2
R
n
OH
R = CH3, C2H5 or

Figure 1: Structures of phenolic polymers synthesized by chemical and enzymatic methods.

availability in pure form at relatively low cost, specificity and activity toward a number of monomers. Horseradish peroxidase (HRP) is active in a number of organic solvents and their mixtures with water [3-5,8]. Phenol, derivatives of phenol and aromatic amines have been polymerized and polymers characterized with the help of a variety of spectroscopic techniques [4]. Phenol polymerization can be carried out in bulk solvents, at the oil-water interface of reversed micelles and at the air-water interface of Langmuir troughs [3,4,7,11]. The polymers exhibit molecular weights ranging from a few hundred to a few tens of thousand depending on the type of monomer and the reaction medium. Significant molecular association of phenolic polymers occurs in solutions of DMF or mixtures of DMF and methanol. This can lead to an overestimation of molecular weights as determined by gel permeation chromatography [5]. Inorganic salts such as LiBr, at sufficiently high concentrations, help dissociate the polymer aggregates.

We have reported earlier on the ability to control the molecular weight and polydispersity of polymers from p-cresol, p-ethylphenol and m-cresol synthesized by peroxidase-catalyzed reactions in reversed micelles and mixtures of DMF or ethanol with water [5,8]. The ability to control polymer molecular weight and polydispersity is desirable since different applications may require different polymer properties that are dependent on molecular weight. One objective of the work has been to produce low molecular weight oligomers and to analyze their application potential in the photoresist industry. Higher molecular weight materials are desirable for applications as abrasives and foundry materials such as machine housings, automotive transmissions and cylinder heads. Polymer processability is another important factor in a number of applications. The hydroxyl groups on the polyphenols prepared by enzymatic methods are amenable to chemical modifications for melt casting, to render them soluble in alkanes for solvent casting or to enhance their UV absorbing characteristics.

We discuss some of our recent results on polymer characteristics and solvent-enzyme-polymer interactions that lead to a molecular level understanding of the phenomena involved in the control of polymer molecular weight.

Experimental

Phenolic monomers and solvents were purchased from Aldrich Chemical Company (Milwaukee, WI). Horseradish peroxidase (Type II) was purchased from Sigma Chemical Company (St. Louis, MO). All chemicals were of highest purity available, and were used as received.

All synthetic and kinetic reactions were carried out at room temperature. A reaction mixture for polymer synthesis was prepared by dissolving 0.1 to 0.2 M monomer in a mixture of DMF or ethanol and buffer. An aliquot of enzyme solution, prepared by dissolving HRP in N-[2-hydroxyethyl]piperazine-N'[2-ethanesulfonic acid] (HEPES) buffer at a known concentration, was added to the monomer solution such that the final enzyme concentration was 0.1 mg/mL. The reaction was initiated by dropwise addition of H_2O_2. The total amount of H_2O_2 added was 30% in excess of the stoichiometric amount relative to monomer. Reactions were continued for 24 hours. For molecular weight analysis, 50 µL of the reaction mixture was diluted with 3 mL of 1% LiBr/DMF solution. The diluted solution, 50 - 100 µL, was injected onto a gel permeation chromatography (GPC) column. Details of the GPC column and conditions for molecular weight analysis were discussed earlier [5]. Polystyrene was used as an external standard.

Experiments to determine the reaction kinetics were carried out with m-cresol and HRP in ethanol/HEPES buffer mixtures of different compositions. The enzyme concentration was reduced to 99 nM (4 µg/mL) in order to obtain slower and measurable reaction rates. m-Cresol concentration was varied between 10 to 50 mM and 5 mM H_2O_2 was used to initiate the reaction. 100 µL samples were taken at a predetermined time, depending on the reaction rate, and diluted with 2.4 mL acetonitrile. Molecular weight analysis was done by GPC with 5-10 µL aliquots of this sample. A 3.9 x 150 mm NOVA-PAK C18 reverse phase column (Waters, Milford, MA) was used to estimate the monomer concentration with the help of a UV detector calibrated at 280 nm. A 56/44 (v/v) acetonitrile/water mixture was used as the eluent at a flow rate of 1 mL/min. Each reaction and injection was run in duplicate. Kinetics were analyzed based on the assumption that the reaction followed pseudo-single substrate kinetics in the presence of excess hydrogen peroxide. However, the concentration of hydrogen peroxide was low enough not to inhibit the enzyme.

Reactions in reversed micellar solutions were carried out as described elsewhere [7]. Polymer at the end of reaction was recovered and purified before analyzing molecular weight. Monomer conversion in the supernatant was chromatographically determined. Thermal analysis on the polymers was carried out in a nitrogen atmosphere at a heating rate of 10°C/min.

Results and discussion

The reaction scheme for HRP-catalyzed phenol polymerization is illustrated in Figure 1. ^{13}C-NMR results from earlier studies showed that the monomer units in poly(p-ethylphenol) are primarily linked at ortho positions as the para position is occupied by the ethyl group. As a result, the polymer is not expected to be cross-linked. On the other hand, m-cresol has three positions for bond formation on the phenylene ring, and the resultant polymer is expected to be highly cross-linked. Thermal characteristics of the polymers support this notion (Figure 2). Differential scanning calorimetry (DSC) thermograms for poly(p-ethylphenol) and poly(m-cresol) show that the latter is thermally stable while the former exhibits a large exotherm at 116°C, an indication of thermal cross-linking in poly(p-ethylphenol) [5]. Thermogravimetric analysis (TGA) showed that the polymers are thermally stable with about 10% weight loss at 300°C in a nitrogen atmosphere.

Details of poly(p-ethylphenol) synthesis and polymer molecular weight and polydispersity control in DMF/water systems have been reported earlier [5]. Results for poly(m-cresol) synthesized in ethanol/water system are presented here. Polymers of p-cresol, p-ethylphenol, p-isopropylphenol and p-butylphenol could also be prepared in ethanol/water mixtures, and the polymers showed a maximum molecular weight of about 2,500. In comparison, poly(m-cresol) could be prepared up to a molecular weight of 25,000. Figure 3 shows the effect of ethanol content in the reaction medium on poly(m-cresol) molecular weight. The data in Figure 3a reflect the molecular weight of the polymer obtained from 0.2 M monomer. Polydispersity in all cases was greater than two. Although the polymer produced in 100% buffer exhibited a molecular weight of ca. 3,000, the polymer yield was extremely poor. The relatively higher molecular weight of poly(m-cresol) compared to para-substituted polyphenols may be attributed to cross-linking in poly(m-cresol). A molecular weight of ca. 7,000 was obtained in 20% ethanol/water mixture wherein the polymer yield was over 90%. Earlier we reported molecular weights as high as 25,000 for poly(m-cresol) at lower monomer concentrations [8]. Lower monomer concentration at a constant enzyme concentration means lower hydrogen peroxide-to-enzyme molar ratio. This leads to minimal enzyme inhibition, higher monomer conversion and polymer yield and higher molecular weight. Although there is no significant difference in the molecular weight of poly(m-cresol) for ethanol content in the reaction medium in the range of 10 to 40%, poly(m-cresol) molecular weight exhibited an overall bell-shaped dependence on ethanol content. This 10-40% ethanol range corresponds well with that for which there is a good-to-significant (50-95%) monomer conversion (Figure 3b). It is therefore logical to attribute this phenomenon to molecular interaction of solvent with enzyme, monomer

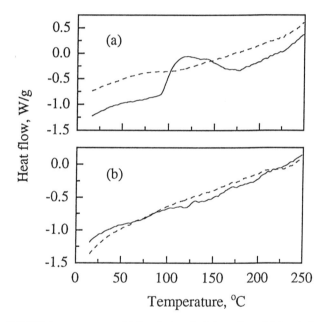

Figure 2: DSC thermograms of (a) poly(*p*-ethylphenol) and (b) poly(*m*-cresol). Solid and dashed lines represent first and second heats, respectively.

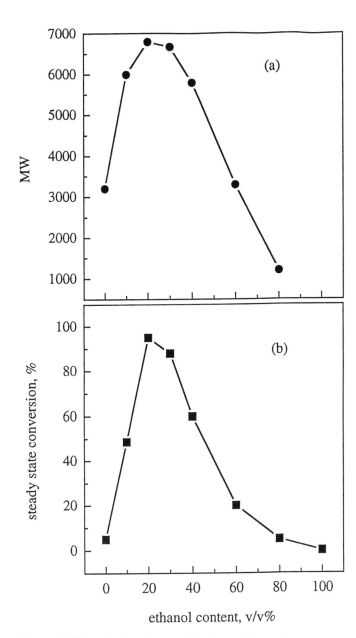

Figure 3: Effect of ethanol content in the reaction medium on (a) poly(*m*-cresol) molecular weight and (b) monomer conversion.

and polymer. While the monomer and polymer are expected to have enhanced solubility at higher ethanol contents, the presence of excess solvent may be detrimental to enzyme activity as shown in a number of solvent-enzyme systems [12]. Therefore, there exists a balance between the solubility of growing polymer chains that leads to enhanced molecular weight and the enzyme activity that is responsible for the monomer conversion. We examined these different molecular interactions that are responsible for the observed poly(m-cresol) molecular weight profile in ethanol-water reaction medium by measuring enzyme activity, by studying the enzyme structure and by estimating the polymer solubility.

Figure 4 illustrates the idealized profiles of enzyme activity and polymer solubility as a function of solvent composition expected from the estimation of solvent-enzyme-polymer interactions that can be experimentally quantified. From the knowledge of solubility parameters for the solvent mixture and the polymer, the solvent-polymer interaction parameter, $\chi_{poly\text{-}solv}$, can be estimated from Flory-Huggins theory. Polymer chains essentially cease to grow after they precipitate out of solution, and keeping the chains solubilized enhances molecular weight. Since ethanol aids polymer solubility, higher molecular weights can be expected at a higher ethanol content in the reaction mixture. However, the enzyme activity is strongly dependent on ethanol content. As the proportion of ethanol increases in the reaction mixture, enzyme activity goes through a maximum. Analogous to dioxane/water and DMF/water systems, the enzyme is completely deactivated at high concentrations of ethanol. However, the enzyme does not appear to be irreversibly deactivated in 100% ethanol. While the enzyme exhibits no activity in 100% ethanol, gradual addition of buffer (to bring ethanol content from 100% to 20%) with all reactants in place leads to near complete restoration of enzyme activity. This observation suggests that the lack of enzyme activity in ethanol may not be completely due to perturbations caused in enzyme structure. Rather, a combined effect of solvent-monomer-polymer-enzyme interactions as a function of solvent composition is reflected in the observed polymer molecular weight profile. Solvent effect on enzyme activity is studied by estimating kinetic constants and by probing the enzyme structure by spectroscopic methods.

Figure 5 shows the Lineweaver-Burk plots for HRP-catalyzed oxidation of m-cresol in a reaction mixture where ethanol content varied from 0% (i.e., all buffer) to 80%. The enzyme was essentially insoluble and inactive in reaction media containing 80 to 100% ethanol. Table I lists the kinetic constants obtained from the plots. The enzyme exhibited most activity (i.e., number of moles of monomer consumed per unit time) in a reaction medium containing 20% ethanol. The Michaelis constant, K_M, decreased continuously as the ethanol concentration increased, which is indicative of the decreasing affinity of the enzyme for the substrate. The apparent loss of enzyme activity may be attributed to the loss of enzyme affinity for the substrate. However, the increasingly weak binding of the monomer to the enzyme may be due to a number of reasons including one or both of the following. (1) The enzyme structure is perturbed so that either the access of substrate to enzyme active site is restricted or the substrate binding is weak; (2) the substrate partitioning between the solvent and the enzyme active site is significantly affected. Ryu and Dordick [13] recognized in their study the role of organic solvents on peroxidase structure and function and reported that monomer-solvent interactions dictate the monomer partitioning behavior between solvent and the enzyme active site. As a result, at high ethanol contents, m-cresol is likely to be overwhelmingly partitioned to the solvent leading to nonavailability of the substrate in the enzyme active site. We are currently probing the enzyme structure as a function of solvent composition using circular dichroism (CD) spectropolarimetry, fluorescence spectroscopy and room temperature electron spin resonance (ESR) spectroscopy. Preliminary results from CD and UV studies indicate that the enzyme secondary structure and heme environment are invariant with solvent composition. For

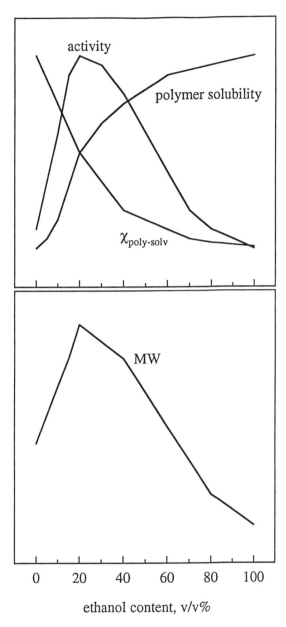

Figure 4: Idealized profiles expected from the estimation of solvent-enzyme-polymer interactions.

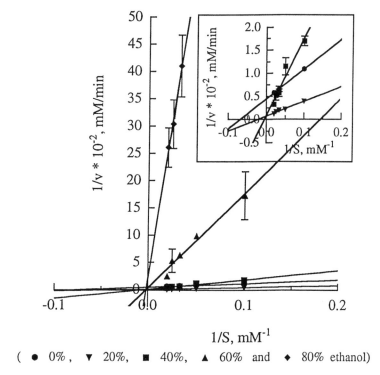

(● 0% , ▼ 20%, ■ 40%, ▲ 60% and ◆ 80% ethanol)

Figure 5: Lineweaver-Burk plots for HRP-catalyzed oxidation of *m*-cresol in a reaction mixture with varying ethanol concentration. (Inset: Expanded region for lower ethanol concentration)

ESR measurements, the enzyme protoheme group is spin labeled [*14*]. This labeling allows us to follow the enzyme active site dynamics at room temperature, which are truly reflective of the kinetic behavior of HRP.

Table I. Kinetic data for HRP-catalyzed oxidation of *m*-cresol in ethanol-water mixtures

Ethanol (% v/v)	V_{max} (mM/min)	K_M (mM)	k_{cat}/K_M $(min*\mu g/mL)^{-1}$
0	221	14	3.9
20	1293	41	7.9
40	1111	187	1.5
60	292	503	0.15
80	47	548	0.02
100	0	-	-

In order to obtain lower molecular weights, *m*-cresol was polymerized in 100% buffer. As seen in Figure 3, the polymer thus prepared exhibited a molecular weight of about 3,000 but the polymer yield was poor. However, this number could be lowered to 1,400 and the monomer conversion could be enhanced to 40% by reducing hydrogen peroxide to enzyme molar ratio. Further, polymer chains could be precipitated as they formed by the addition of small amounts of salt, and the polymer could be recovered by filtration. Monomer conversion could also be enhanced by pulsed addition of the enzyme. A process was developed to produce polymer with control of molecular weight and to enhance monomer conversion (Figure 6). The ethanol content in the feed and the residence time in the reactor were adjusted to obtain a desired polymer molecular weight. Enzyme solution and H_2O_2 were added in pulses to optimize the utilization of the enzyme. The supernatant, containing unreacted monomer, hydrogen peroxide and enzyme, was recycled after isolating the polymer.

FTIR studies on polyphenols prepared by enzymatic processes indicated that the polymers lacked ether links and that the hydroxyl groups were intact. This fact allows derivatization of hydroxyl groups with a number of substituents to enhance the polymer functional properties. For example, the hydroxyl groups on poly(*p*-ethylphenol) were esterified with palmitoyl chloride and cinnamoyl chloride in DMF in the presence of stoichiometric amounts of pyridine [*5*]. The derivatized polymers exhibited lower melting points and/or enhanced UV absorption characteristics. FTIR spectra, shown in Figure 7 for palmitic groups substituted on the polymer backbone, revealed esterification of nearly all hydroxyl groups. In addition, the degree of esterification could be controlled by adjusting the stoichiometry of the reactants. The polymer may also be functionalized with biotin groups to bind a number of biomolecules to the polymer matrix. Cell growth factors, such as arginine-glycine-aspartic acid (RGD) tripeptide, may be attached to the polyphenol backbone.

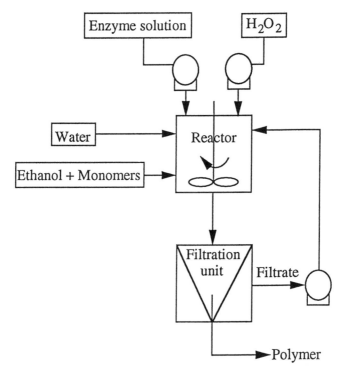

Figure 6: A schematic of the process for the production of phenolic polymers with control of polymer molecular weight and monomer conversion.

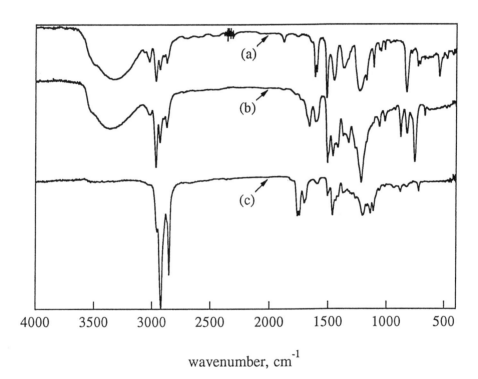

4000   3500   3000   2500   2000   1500   1000   500

wavenumber, cm^{-1}

Figure 7: FTIR spectra for (a) *p*-ethylphenol, (b) poly(*p*-ethylphenol) and (c) poly(*p*-ethylphenol) derivatized at hydroxyl groups by palmitoyl chloride.

Conclusions

The molecular weight of poly(m-cresol) could be controlled in ethanol/water mixtures by manipulating the composition of the reaction medium. Polymer precipitation could be induced in the medium with the help of salts to curtail further chain growth. A process scheme was developed to produce polymer on a large scale while controlling the molecular weight. In addition, monomer conversion could be enhanced while recycling the reaction medium. The polymer was thermally stable, and could be derivatized at the available hydroxyl groups to tailor functional properties. In order to understand different molecular interactions in the reaction medium that are responsible for the observed dependence of polymer molecular weight on the solvent composition, enzyme activity as a function of the reaction medium composition was studied. Preliminary results indicated that the enzyme activity loss at high ethanol contents could be significantly influenced by the partitioning behavior of the monomer. Further studies on solvent-mediated alterations in enzyme structure are currently underway to gain further insight into these effects.

Acknowledgment

This work was performed while MA held a National Research Council-NRDEC Research Associateship.

Literature cited

1. Brode, G.L., *Kirk-Othmer Encyclopedia of Chemical Technology*; John Wiley & Sons: New York, 1982; Vol. 17, pp. 384-416.
2. White, D.M.; Cooper, G.D. *Kirk-Othmer Encyclopedia of Chemical Technology*; John Wiley & Sons: New York, NY, 1982; Vol. 18, pp. 594-615.
3. Dordick, J.S.; Marletta, M.A.; Klibanov, A.M. *Biotechnol. Bioeng.*, **1987**, 30, pp 31-36.
4. Akkara, J.A.; Senecal, K.J.; Kaplan, D.L. *J. Polym. Sci.*, **1991**, 29, pp 1561-1574.
5. Ayyagari, M.; Marx, K.A.; Tripathy, S.K.; Akkara, J.A.; Kaplan, D.L.; *Macromolecules*, **1995**, 28, pp 5192-5197.
6. Liang, R-C., Pokora, A.R.; Cyrus, W.L. U.S. Pat. 5,212,044, **1993**.
7. Rao, A.M.; John, V.T.; Gonzalez, R.D.; Akkara, J.A.; Kaplan, D.L.; *Biotechnol. Bioeng.*, **1993**, 41, pp 531-540.
8. Ayyagari, M.; Akkara, J.A.; Kaplan, D.L.; *Acta Polym.*, **1996**, 47, pp 193-203.
9. Weininger, S.J.; Steirmitz, F.R. *Organic Chemistry*, Academic Press: New York, NY, 1984.
10. Westerfield, W.W.; Lowe, C. *J. Biol. Chem.*, **1942**, 145, pp 463-470.
11. Akkara, J.A.; Kaplan, D.L.; Samuelson, L.A.; Bruno, F.F.; Mandal, B.K.; Marx, K.A.; Tripathy, S.K.; U.S. Pat. 5,143,828, **1992**.
12. Kazandjian, R.Z.; Dordick, J.S.; Klibanov, A.M. *Biotechnol. Bioeng.*, **1986**, 28, pp 417-421.
13. Ryu, K.; Dordick, J.S. *Biochemistry*, **1992**, 31, pp 2588-2598.
14. Asakura, T.; Leigh, J.S.; Drott, H.R.; Yonetani, T.; Chance, B. *Proc. Nat. Acad. Sci.*, **1971**, 68, pp 861-865.

Chapter 7

A Biocatalytic Approach to Novel Phenolic Polymers and Their Composites

Sukanta Banerjee[1], Premchandran Ramannair[1], Katherine Wu[1], Vijay T. John[1], Gary Mcpherson[2], Joseph A. Akkara[3], and David Kaplan[3]

Departments of [1]Chemical Engineering and [2]Chemistry, Tulane University, New Orleans, LA 70118
[3]Biotechnology Division, U.S. Army Soldier Systems Command, Natick Research, Development, and Engineering Center, Natick, MA 01760–5020

We describe research to understand basic structure/function properties of conjugated phenolic polymers and their composites, and to exploit applications in enzyme delivery systems, coatings technologies and luminescent materials. The polymers are synthesized enzymatically, and the reaction is very feasible when carried out in the essentially nonaqueous system of reversed micelles. The reaction is catalyzed by horseradish peroxidase, which fully retains catalytic activity in these systems. Additionally, monomer solubility is significantly enhanced through hydrogen bonding with the surfactant. The polymer is produced in a microspherical morphology and the internal density of these microspheres can be controlled. It is possible to encapsulate enzymes in these microspheres leading to novel enzyme delivery systems. Depending on the monomer used, it is possible to prepare luminescent polymers, and polymers that bind to semiconductor nanoparticles.

The use of enzymes to synthesize novel polymers is an interesting and potentially very useful concept. There is a biomimetic aspect associated with such synthesis, the process is usually environmentally benign, and the material often has novel properties not easily achievable through chemical synthesis. We consider the synthesis of polyphenolics using an oxidative enzyme such as horseradish peroxidase, a reaction that has a mechanistic analogy in the synthesis of lignin (1). In addition to the environmentally benign aspect of synthesizing polyphenols for conventional coatings applications without the use of formaldehyde as a reaction intermediate, the enzymatic route to polyphenol synthesis leads to conjugated polymers. The feasibility of enzymatic polphenol synthesis in organic solvents was demonstrated by Dordick at el. (2) and structural characterizations of both polyphenols and poly(aromatic amines) were done by Akkara et al. (3). The synthesis is illustrated through the simplified mechanism shown in Figure 1. Phenoxy radical centers initially formed on the monomer or growing chain migrate to the ortho positions (the para position while mechanistically allowed, is less favored) following which, coupling through condensation occurs. The direct ring-

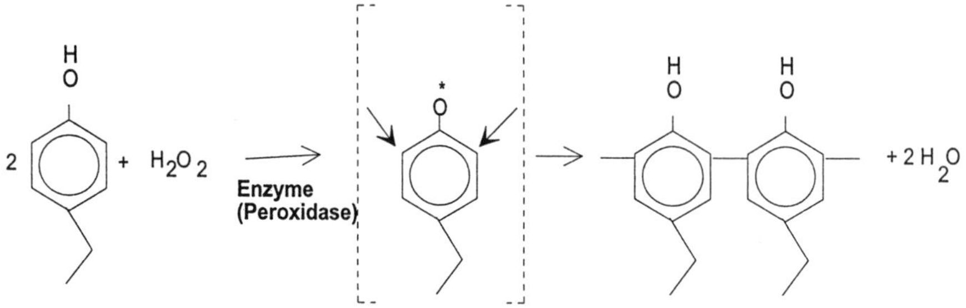

Figure 1. Simplified schematic of the polymerization reaction. The bold arrows (in the phenoxy radical) indicate coupling in the ortho positions. The resonance structures also imply the possibility of para coupling.

to-ring coupling leads to the formation of conjugated polymers with potential applications to nonlinear optics (*4*).

In this paper, we review our recent research on the enzymatic synthesis of polyphenols, with a view to developing some new applications. The novelty of our method of enzymatic polymer synthesis is that we carry out the reaction in surfactant-based microstructured environments, where the system microstructure is the result of surfactant and monomer self-assembly. In particular, we have studied the environment of reversed micelles. These are water-in-oil microemulsions stabilized by surfactant. The double-tailed anionic bis(2-ethylhexyl) sodium sulfosuccinate, or AOT, shown in Figure 2a, is very effective in forming these microstructured fluid phases. Reversed micelles are capable of solubilizing macromolecules, including proteins, in the microaqueous core. Enzymes solubilized in reversed micelles are catalytically active in what is an essentially nonaqueous system. The rationale for polymer synthesis in this environment is the following. Phenolic monomers have limited solubility in water, and chain growth may be drastically retarded due to chain insolubility if the reaction is carried out in aqueous media. On the other hand, the tuning of monophasic organic solvents to maintain enzyme activity and sustain polymer growth is not always a simple task. The reversed micelle environment is an effective compromise. The enzyme is catalytically very active when solubilized in the water core, and the organic bulk phase helps support monomer and chain solubilization. A second interesting aspect of synthesis in reversed micelles is the partitioning of the polar monomer to the oil-water interface. This partitioning is the result of surfactant-phenol hydrogen bonding interactions. In dry (water-free) reversed micelles, such interactions lead to a dramatic phase transition to an organogel (*5*). The partitioning to the interface (depicted by the arrow of Figure 2b with the head of the arrow representing the hydroxyl moieties of the monomer) is clearly indicated by perturbations to the surfactant C=O as shown for the monomers 2-naphthol and 4-ethylphenol in Figure 3. The partitioning may result in a prealignment of the monomers before synthesis, and also provides a means of monomer replenishment to the vicinity of the enzyme. Additionally, such surfactant-monomer hydrogen bonding interactions significantly enhance monomer solubility in the reaction medium.

Polymerization in reversed micelles is very easy to carry out, and for brevity, the reader is requested to our earlier papers for details on procedures (*6*)(*7*). The self-assembling nature of the system implies that adding the enzyme dissolved in buffer (HEPES), to the surfactant and the monomer dissolved in the solvent (isooctane), leads to spontaneous formation of reversed micelles with encapsulated enzyme, and monomer partitioning to the micellar interface. Reaction is initiated through the addition of H_2O_2 in small aliquots. The reaction is rapid and within 5-10 minutes of reaction initiation, over 80% of the monomer becomes converted to polymer which precipitates out of solution. An interesting aspect of the reaction conducted in reversed micelles is the observation that the polymer precipitates out in the morphology of microspheres as the scanning electron micrograph of Figure 4 illustrates. The microsphere morphology can be reproducibly obtained when the surfactant to monomer molar ratio in the micellar system is at least 3/1 (*7*).

This paper has a focus on two aspects of polyphenol synthesis in reversed

Bis(2-ethylhexyl) sodium sulfosuccinate (AOT)

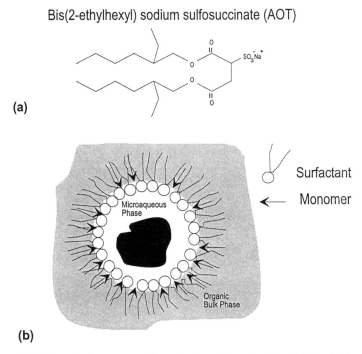

(a)

(b)

Figure 2. (a) Chemical structure of the anionic surfactant sodium bis(2-ethylhexyl) sulfosuccinate (b) Schematic of enzyme solubilized in the micelle and monomer partitioning to the micelle interface. The arrow here refers to the monomer.

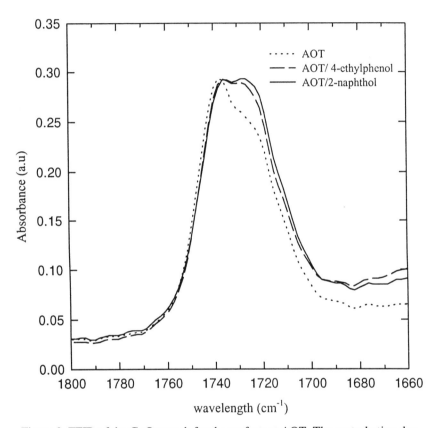

Figure 3. FTIR of the C=O stretch for the surfactant AOT. The perturbation due to hydrogen bonding with 2-naphthol is noted.

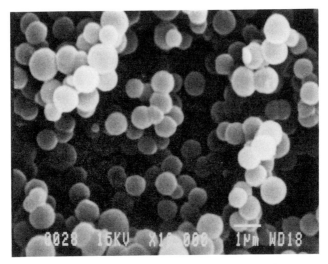

1 μm

Figure 4. Scanning electron micrograph of poly(4-ethylphenol) prepared in the reversed micelle solution.

micelles. The first aspect is a study of the functional properties of such polymers, in particular, the concept of synthesizing fluorescent polymers and polymer-semiconductor nanocomposites. The second aspect is based on polymer morphology, where the microsphere morphology is exploited for specific applications in coatings, enzyme delivery and nanocomposite technology. The paper is essentially a summary of our recent findings with suggestions for future opportunities in this area.

The Enzymatic Synthesis of Functional Phenolic Polymers with Luminescent Properties

Polymers with Luminescent Chromophores. Luminescent polymers have applications in technologies related to plastic scintillators, luminescent solar concentrators, high laser resistant materials, laser dyes and fiber optic sensors (8). The naphthol chromophore for example, can be used in the synthesis of polynaphthols. The enzymatic synthesis of naphthol based polymers in organic solvents has been reported (2,3) but no study has been carried out of its fluorescent properties. Poly(2-naphthol) is very efficiently synthesized in reversed micelles with polymer yields exceeding 90%. The fluorescence characteristics of the polymer are shown in Figure 5. On excitation at 327 nm (the excitation maximum for the monomer), the polymer exhibits an emission spectra with an intense peak at 375 nm and two other peaks of much lower intensity centered around 455 and 481 nm. The monomer emits at 355 nm. Thus, the band at 375 nm is assigned to emission from naphthol residues in the polymer with the 20 nm red-shift from the monomer to the polymer attributed to the increase in conjugation. However, the emission intensities at 455 nm and 481 nm get significantly stronger when the polymer is excited at 413 nm. Excitation at 413 nm reveals a well-structured and highly reproducible emission with intense peaks at 455 nm and 481 nm. These bands cannot be assigned to naphthol units and the bands represent emitting states distinct from the naphthol π-π* singlet state. Excimer/exciplex based emissions are ruled out since such complexes would yield broad structureless bands independent of the excitation wavelength. Our hypothesis is that the the emission at these higher wavelengths results from a more conjugated chromophore bound within the polymer. Mechanistically such chromophores can be generated through coupling at resonance positions besides the 1-1' mode. For example, 6-6' coupling followed by further oxidation to the quinonoid form shown in the inset of Figure 5 leads to a rigid, conjugated chromophore. Infrared spectroscopy points out the the possible existence of quinonoid carbonyl peaks, and NMR evidence points to coupling modes besides the 1-1' positions. Further evidence comes from studies of the electrochemical oxidation of 2-naphthol indicating the possible formation of quinonoid segments (9). The rigid, planar, conjugated structure of the quinonoid segment indicates the possibility that these are the fluorophores giving rise to the emission with features at 455 and 481 nm. Since fluorescence is sensitive to trace concentrations of highly fluorescent groups in a polymer backbone when appropriately excited, it is possible that even small amounts of these fluorophores may be responsible for the structured emission at 455 and 481 nm. Figure 6 illustrates the potential of tuning fluorescence using polymers (and copolymers) from 2-naphthol and 1-hydroxypyrene. A full study of the nature of this fluorescence to

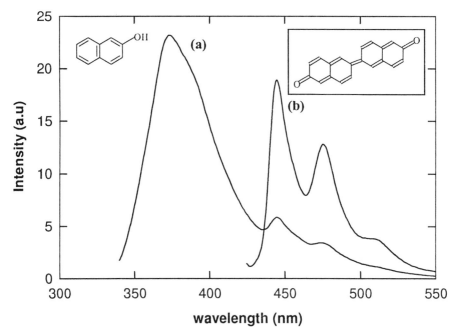

Figure 5. Fluorescence characteristics of poly(2-naphthol). Trace (a) is the emission spectrum when the polymer is excited at 327 nm, trace (b) the emission spectrum when excited at 413 nm. The second emission is tentatively attributed to the quinonoid moieties shown in the inset.

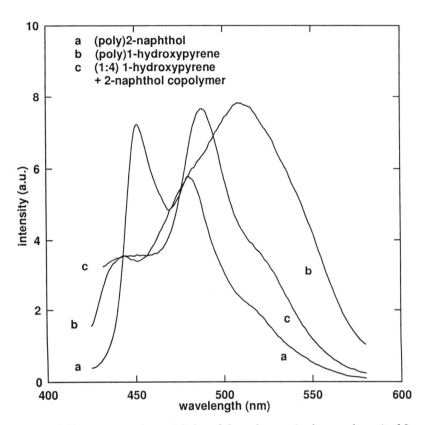

Figure 6. Fluorescence characteristics of the polymers (and a copolymer) of 2-naphthol and 1-hydroxypyrene. Excitation at 413 nm.

correlate structure to fluorescence, has not been completed. However, we note the significant Stokes shift on exciting at 413 nm, the polymer from 1-hydroxypyrene.

Polymer-Semiconductor Nanocomposites. A second approach to building polymers with novel optical properties is to bind semiconductor nanoparticles such as cadmium sulfide to the polymers. Such polymer-semiconductor nanocomposites possess the optoelectronic properties of the semiconductor material, and additionally may provide greater flexibility in processing. The sulfur atom of thiol compounds binds strongly to surface cadmium ions of CdS and CdSe semiconductor nanoparticles. Such surface passivation has been shown to reduce nonradiative recombinations leading to better luminescence properties (*10*). The objective in our work has been to use a thiol-containing monomer to prepare polymer-CdS nanocomposites. Thus, the monomer 4-hydroxythiophenol can be used for the preparation of such nanocomposites with the thiol groups on the polymer bound to CdS and CdSe nanoparticles. Again, the reversed micellar system is ideal for such synthesis. The microenvironment of the water pools restricts growth of inorganic clusters to the nanometer size range. Nanoparticle size can be controlled simply by adjusting micelle size, which in turn is directly related to the system water content (*11*)(the quantity, w_0, the water to AOT molar ratio is a measure of micelle size). The micellar environment allows CdS to be synthesized with bandgap values varying from the bulk material (2.4 eV) up to about 3 eV with particle sizes approaching 2 nm (*12*).

Although both CdS and 4-hydroxythiophenol containing polymers can be synthesized in reversed micelles, there are some nuances to the procedure. First, since Cd^{2+} rapidly deactivates HRP, the CdS synthesis has to be done separately from the enzymatic polymer synthesis. The procedure that we have adopted is to first synthesize 4-hydroxythiophenol containing polymers and then contact the polymer with CdS nanoparticles. When reaction in reversed micelles is carried out with pure 4-hydroxythiophenol, the product is an insoluble precipitate that is hard to process. This is probably due to H_2O_2 induced oxidation of thiol groups leading to the formation of disulfide bonds, which gives rise to a highly crosslinked, unprocessible material. However, copolymers of 4-hydroxythiophenol and 4-ethylphenol are very easy to synthesize, and the product shows retention of thiol groups (FTIR analysis). After synthesis, the copolymer is dissolved in dimethylsulfoxide and saturated with CdS nanoparticles. Figure 7 illustrates fluorescence characteristics of these copolymers. It is seen that as the fraction of 4-hydroxythiophenol in the copolymer increases, the fluorescence intensity increases indicating a greater saturation of CdS in the polymer. Atomic absorption analysis of polymer cadmium content indicates an increase from 3 wt% cadmium in copoly(10% 4-hydroxythiophenol, 90% 4-ethylphenol) to 7.2 wt% cadmium in copoly(50% 4-hydroxythiophenol, 50% 4-ethylphenol). An interesting aspect of the fluorescence is that nanoparticle capping by the polymer significantly cuts off low energy surface recombinations (which usually show up around 540 nm); the emission observed corresponds to recombinations with near bandgap energies. The polymer-capped CdS nanocomposite is remarkably stable in solution as opposed to colloidal CdS which precipitates out due to Ostwald ripening and photocorrodes upon exposure to light and oxygen (*13*). We note that up to about 60% 4-hydroxythiophenol

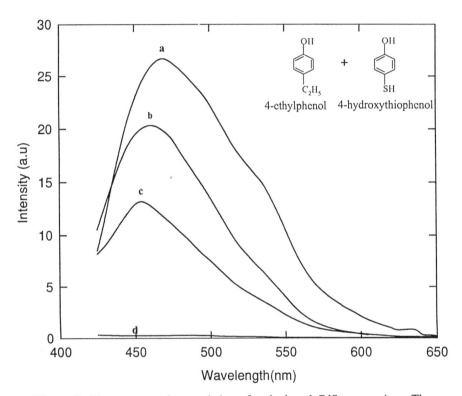

Figure 7. Fluorescence characteristics of polyphenol-CdS composites. The monomer ratios used to make the of the various copolymers (a) 1/1 4-ethylphenol (EP)/ 4-hydroxythiophenol (HTP), (b) 7/3 EP/HTP, (c) 9/1 EP/HTP. Curve (d) is the polymer of (b) but without CdS. As the HTP content increases, the amount of CdS bound to the polymer also increases as shown by the intensity enhancement.

in the monomer mixture, the copolymer precipitates out from reversed micelles in the microsphere morphology of Figure 3, and can be easily dissolved in a number of polar solvents. At higher levels of the 4-hydroxythiophenol component, the morphology is lost and the precipitate becomes unprocessible, perhaps due to the prevalence of disulfide bonds.

Applications resulting from polymer morphology

Two important aspects result from the the the observation that the enzymatic route to polyphenol synthesis in reversed micelles leads to polymers with the morphology of microspheres.

1. The internal density of the microspheres can be controlled by adjustment of reaction time and H_2O_2 addition (14). This is so because the microspheres contain enzyme and unreacted monomer, and polymerization continues even after the overall morphology has been achieved, leading to densification. Figure 8 illustrates the densification of the microspheres as observed through transmission electron microscopy. The diffuse microspheres of Figure 8a form within 5-15 minutes of reaction initiation, while continuing reaction over 24 hours results in the dense microspheres of Figure 8b. By stopping reaction after the first 15 minutes and washing the polymer to remove residual monomer and H_2O_2, it is possible to arrest densification.

2. During precipitation, the polymer also encapsulates solutes located in the water core of the micelle. This may be a novel method, perhaps a ship-in-a-bottle approach to microencapsulation. The solutes include both organic water-soluble macromolecules such as proteins, and inorganic nanoclusters.

Let us consider two examples of such microencapsulation. In the first example, ferrite nanoparticles are synthesized in the micellar environment with a size of the order of magnetic domain size. The particles then have room temperature superparamagnetic properties. After particle synthesis, monomer and enzyme are added to the micellar system and polymerization conducted. The interfacial nature of polymerization results in an encapsulation of the ferrite particles in the polymer chains, the final polymer precipitates contain ferrite nanoparticles. Figure 9 illustrates a cut-section TEM of a single polymer particles with the dark spots being ferrite nanoparticles. These composites exhibit superparamagnetism (15) and being in the form of microspheres can be easily dispersed for magnetic coatings applications. Additional applications include the use of these composites for magnetic bioseparations since phenolic materials can be easily functionalized to bind to specific antibodies (16). The use of these magnetic microspheres as contrast agents in magnetic resonance imaging is another potential application.

The second example is on enzyme encapsulation. Here, other enzymes are cosolublized with HRP in the reversed micelle system. When polymerization is carried out, the precipitating polymer not only traps HRP but also the cosolublized enzyme. Short duration polymerization to maintain the diffuse internal density of the microspheres, results in polymer-enzyme composites that retain the catalytic activity of the entrapped enzyme (accomodating for diffusional limitations). For example, we have shown that phosphodiesterase entrapped in the microspheres is catalytically active and has an enhanced stability compared to the free enzyme (13).

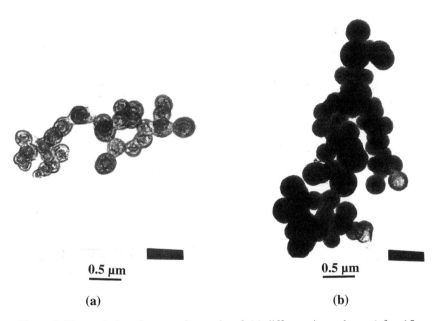

(a)   **(b)**

Figure 8. Transmission electron micrograhs of (a) diffuse microspheres (after 15 mins reaction) and (b) dense microspheres (after 24 hours reaction).

However, activity retention is also dependent on the ability of the enzyme to survive the oxidative addition of H_2O_2 to initiate polymerization. Metalloenzymes may not always be good candidates for microencapsulation through this method, due to some H_2O_2 induced deactivation. To circumvent the problem, we have carried out a two step process of metalloenzyme microencapsulation (*17*). The phenolic polymer is first synthesized in reversed micelles using HRP. Subsequently the polymer is dissolved at high concentrations in a solvent (acetone, or benzene). The polymer solution is then contacted with a much larger volume of a reversed micellar solution containing the enzyme of interest. Since the reversed micellar solution is a nonsolvent, the polymer reprecipitates. Interestingly the precipitate has a microsphere morphology. As a possible consequence of surfactant-polymer interactions, we find that the enzyme cosolubilized in the micelles is again entrapped in the polymer microspheres with high efficiency (*16*). Figure 10 illustrates an interesting hollow microsphere morphology upon polymer precipitation in the micellar nonsolvents. In recent research, we have encapsulated the phosphotriestase from *Pseudomonas diminuta* into these microspheres. The enzyme is capable of hydrolyzing organophosphorus compounds including nerve agents (*18*). While the free enzyme has a turnover frequency of about 3000 s^{-1} (k_{cat} values) for the hydrolysis of paraoxon, we have been able to approach values of 500 s^{-1} with the encapsulated enzyme. The Lineweaver-Burk plot of Figure 11 reports the Michaelis-Menten parameters for the hydrolysis of paraoxon. The lyophilized polymer-enzyme composite has good shelf life. The results indicate the viability of the composite in technologies related to chemical decontamination. The catalytic microspheres may be used to make protective coatings on fabrics. Or they may be sprayed onto the site of chemical decontamination. As a result of encapsulation, the enzyme may be protected from shear-induced denaturation upon spraying. Studies are in progress to test the efficacy of the composite after being subjected to shear, and in the presence of foams.

Summary and Future Directions

The results described indicate the potential of using enzymes to couple phenols to produce functionally useful polymers, with particular emphasis on polymers and polymer-semiconductor composites with luminescent properties. The possibility of deriving fluorescence either from the polymeric component or from the semiconductor nanoparticle adds considerable flexibility to the synthesis scheme. The fact that these are conjugated polymers implies the possibility of electroluminescence. These materials may therefore represent a new class of polymeric electronic materials, and polymeric photocatalysts, with electronically coupled quantum dot clusters (*19*) Continuing studies seek to clarify these issues and to delineate new applications. Dual application possibilities also exist with enzymatically coupled dihydroxynaphthalenes. The recent, remarkable report by Dordick and coworkers (*20*) indicates an indirect approach to poly(hydroquinone) synthesis using a multienzyme, chemoenzymatic method to avoid direct peroxidase-catalyzed oxidation of hydroquinone to benzoquinones. The resulting polymer is redox active and has many electrochemical applications in battery and sensor technologies. Polymers from 2,6-dihydroxynaphthalene may have both redox activity and fluorescence properties, widening their applications to chemical sensors.

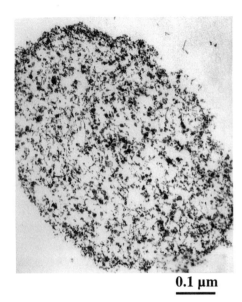

0.1 μm

Figure 9. A cut section transmission electron micrograph of ferrite nanoparticles embedded in polyphenol microspheres. The elliptical cross-section is due to distortion during sectioning.

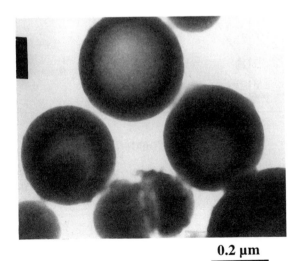

0.2 μm

Figure 10. Polymer precipitation using a micellar nonsolvent. The TEM indicates partially hollow microspheres formed as a result of phase segregation between incorporated water in the microsphere and the bulk organic phase.

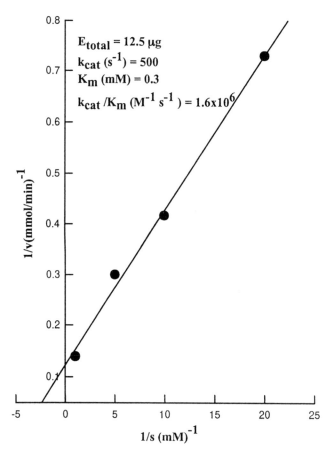

Figure 11. Lineweaver-Burk plot of phosphotriesterase activity when encapsulated in polymer microspheres. The reaction is the hydrolysis of paraoxon (diethyl *p*-nitrophenylphosphate). Nitrophenol formation is monitored by absorbance at 410 nm.

The nature of synthesis in reversed micelles and surfactant microstructures provides another interesting aspect meriting further study. We have shown that the reversed micellar environment influences polymer formation in the morphology of microspheres whose internal density can be controlled. The mechanism of particle formation and polymer precipitation also involves encapsulation of intramicellar solutes. Polyphenol precipitation in a micellar nonsolvent is an alternate method of microencapsulation. Our results have shown the potential for preparing polymeric microspheres containing active enzymes, or inorganic materials. An extension would be the development of polymer-polymer composites where polyphenols are folded around other polymeric nanoparticles synthesized in reversed micelles.

Finally, the remarkable nature of the enzymatic polymer synthesis in microstructured environments is demonstrated through some very recent results illustrated in the SEM of Figure 12. Here we synthesize poly(4-ethylphenol) in a hard, clear gel made with lecithin (phosphatidylcholine), AOT, water and isooctane. The species concentrations are 0. 2M (lecithin), 0.4 M (AOT) and w_0 ($[H_2O]/[AOT]$) = 70. The composition of the gel is such that the water and isooctane volumes used are almost equivalent, so that the gel is neither an aqueous gel nor an organogel. The gel is formed by repeated warming and cooling and the formation occurs over a period of a few hours. The gel is an extension of the lecithin organogels formed by Luisi and coworkers (*21*) with the difference being the inclusion of AOT as an additional component, and a much larger water content. We have found that the gels are also formed easily in crude lecithin preparations (Sigma Type IV-S, from soybean, for example) which are much less expensive than purified lecithin. The objective of doing polymerization in these gels was to conduct the reaction in a medium which would prevent phase separation of the polymer from the monomer, and thereby allow continued growth. The aqueous gel microphase is used to sustain the enzyme, while the organic gel microphase helps solubilize the monomer. As polymerization proceeds, the polymer can be visually observed growing throughout the gel phase (as a deep grey color spreading through the medium). The polymer formed in this environment was imaged by simply dipping a microscope stub (face-down) into the gel, followed by gentle washing with isooctane to remove the adsorbed AOT and lecithin. As Figure 12 indicates, there is a furrow-like pattern to the polymer. The furrows appear to be made up of coalesced microspheres. It is unclear whether this pattern formation is the result of interfacial polymerization, or of polymerization in the restricted geometry of the organic phase which sustains the monomer. Nevertheless, the results indicate new and fascinating approaches to polymer templating. It may be possible to grow inorganic materials in the channels between the polymer strings, leading to the formation of new composites. There are many such aspects of polymer and composite synthesis in surfactant based microstructures that deserve further study.

Acknowledgement
Financial support by the U.S. Army, the National Science Foundation, and the Tulane/DoD Center for Bioenvironmental Research is gratefully acknowledged. We are grateful to Dr. Frank Raushel (Texas A&M University) for a sample of phosphotriesterase from *Pseudomonas diminuta*.

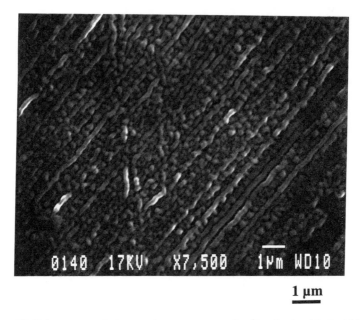

1 μm

Figure 12. Polymer morphology and arrangement when synthesized in lecithin + AOT gels.

Safety Considerations

Cadmium compounds and hydroxythiophenol are toxic and should be handled with appropriate precautions. The MSDS sheets on all organic compounds should be consulted before use.

References

1. B. Halliwell and J.M.C. Gutteridge, *Free Radicals in Biology and Medicine*, Clarendon Press, Oxford, 1989.
2. Dordick, J.; Marletta, M.A.; Klibanov, A.M. *Biotechnol. Bioeng.* **1987**, *30*, 31.
3. Akkara, J.A., Senecal, K.J.; Kaplan, D.L. *J. Polym. Sci.* **1991**, *29*, 1561.
4. Akkara, J.A.; Ayyagari, M.; Bruno, F.; Samuelson, L.; John, V.T.; Karayigitoglu, C.; Tripathy, S.; Marx, K.A.; Rao, D.V.G.L.N.; Kaplan, D.L. *Biomimetics* **1994**, *2*, 331.
5. Tata, M.; John, V.T.; Waguespack, Y.Y.; McPherson, G.L. *J. Phys. Chem.* **1994**, *98*, 3809.
6. Rao, A.M.; John, V.T.; Gonzalez, R.D.; Akkara, J.A.; Kaplan, D.L. *Biotechnol. Bioeng.* **1993**, *41*, 531.
7. Karayigitoglu, C.; Kommareddi, N.; John, V.; McPherson, G.; Akkara, J.; Kaplan, D. *Materials Science and Engineering. C: Biomimetic Materials, Sensors and Systems* **1995**, *2*, 165.
8. Barashkov, N.N.; Gunder, O.A. *Fluorescent Polymers*, Elis Horwood Series in Polymer Science and Technology: New York, **1994**.
9. Pham, P.C.; Lacaze, P.C.; Genoud, F.; Dao, L.H.; Nguyen, M. *J. Electrochem. Soc.* **1993**, *140*, 912.
10. Majetich, S.A.; Carter, A.C. *J. Phys. Chem.* **1993**, *97*, 8727.
11. Pileni, M.P.; Zemb, T.; Petit, C. *Chem. Phys. Lett.* **1985**, *118*, 4.
12. Henglein, A. *Chem. Rev.* **1989**, *89*, 1861.
13. Robinson, B.H.; Towey, T.F.; Zourab, S.; Visser, A.J.W.B.; van Hoek, A. *Colloids Surf.* **1991**, *61*, 175.
14. Xu, X.; Kommareddi, N.; McCormick, M.; Baumgartner, T.; John, V.T.; McPherson, G.L.; Akkara, J. A.; Kaplan, D.L.; *Materials Science and Engineering C: Biomimetic Materials, Sensors and Systems* **1996**, *4*, 161.
15. Kommareddi, N.S.; Tata, M.; John, V.T.; McPherson, G.L.; Herman, M.; Lee, Y.S.; O'Connor, C.J.; Akkara, J.A.; Kaplan, D.L. *Chemistry of Materials*, **1996**, *8*, 801.
16. Nustad, K.; Danielson, H.; Rieth, A.; Funderud, S.; Lea, T.; Vartdal, F.; Ugelstad, J. In *Microspheres: Medical and Biological Applications*, Rembaum, A., Tokes, Z.A., Eds.: CRC Press: Boca Raton, FL, 1988: p 53.
17. Banerjee,S.; Premchandran, R.; Tata, M.; John,V.; McPherson, G.; Akkara, J.; Kaplan, D. *Ind. Eng. Chem. Research* **1996**,*35*, 3100.
18. Dumas, D.P.; Caldwell, R.; Wild, J.R.; Raushel, F.M. *J. Biol. Chem.* **1989**, *264*, 19659.
19. Noglik, H.; Pietro, W.J. *Chem. Mater.*, **1995**, *7*, 1333.
20. Wang, N.; Martin, B.D.; Parida, S.; Rethwisch, D.G.; Dordick, J.S. *J. Amer. Chem. Soc.*, **1995**, *117*, 12885.
21. Capitani, D.; Rossi, E.; Segre, A.L.; Giustini, M.; Luisi, P.L. *Langmuir*, **1993**, *9*, 685.

Chapter 8

Hydrolysis of the Dimethyl Ester of N-Succinylphenylalanine, a Model of Polyesteramides, in the Presence of Papain

Kinetic Study and Computer Simulation

C. David[1], F. G. O. Lefebvre[1], R. Brasseur[3], and M. Vanhaelen[2]

[1]Chimie des Polymères et des Systèmes Organisés, Université Libre de Bruxelles, Campus Plaine, C.P. 206–1, Boulevard du Triomphe, B–1050 Brussels, Belgium
[2]Laboratoire de Pharmacologie et de Bromatologie, Université Libre de Bruxelles, C.P. 205–4, Boulevard du Triomphe, B–1050 Brussels, Belgium
[3]Centre de Biophysique Moléculaire Numérique, Faculté Universitaire de Gembloux, Passage des Déportés, 2, B–5030 Gembloux, Belgium

There is a need to understand the kinetics and mechanism for hydrolysis of condensation polymers such as polyesters and polyamides, both for developing degradable polymers and for the development of enzymatic synthesis. Although kinetics and mechanisms have been studied to some extent for poly-β-hydroxyalkanoates and polycaprolactone, many issues remain to be resolved and require further attention. One of these is the interaction between enzymes and substrate to be hydrolyzed. This work concerns kinetics of the enzymatic hydrolysis of polyesteramides using model compounds, the L- and D-dimethylester of N-succinyl-phenylalanine, with papain. Computer simulation has been used to obtain the conformation, H-bonding and potential energy of the substrates in the Michaelis complex involved in the reaction. Kinetic results are discussed in relation with the computed data.

Biodegradable polymers could gain a broad use in the next few years either as low cost materials to be used for packaging, sanitary and agricultural uses or as more expensive materials for medical uses. For biomedical applications, polyester-amides and polyester-urea have been prepared to combine degradability of polyester with hydrogen bonding ability of the amide groups. The kinetics and mechanisms of ester hydrolysis has been broadly studied, mainly in the case of poly-β-hydroxybutyrate and polycaprolactone. Many problems related to the mechanisms of hydrolysis remain however unsolved. One of them is the nature of interactions between enzymes and microorganisms with the substrate to be hydrolyzed. Many examples concern interactions between enzymes and soluble substrates. The case of insoluble substrates is far less known although recent work has been devoted to the hydrolysis of cellulose by cellulases (*1*), (*2*). Our work is concerned with the kinetics of enzymatic hydrolysis of substrates containing ester and amide groups and with the interactions between the active site of the enzyme and the function to be hydrolyzed. Special attention is paid to differences between a soluble substrate which can fit the active site and a crystalline or semi-crystalline substrate. The following substrate is used in the present study:

$$CH_3O-CO-CH-NH-CO-CH_2-CH_2-CO-OCH_3$$
$$|$$
$$CH_2$$

This dimethylester of N-succinylphenylalanine presents two esters and one amide group and is derived from either L or D phenylalanine. The enzyme is papain, a thiol protease obtained from the fruit of *Carica papaya* which possesses both esterase and amidase activity. The structure of the enzyme has been determined by X-ray diffraction (*3*), allowing easier simulation of the structure and energetic properties of the enzyme-substrate complex.

The Mechanism of Hydrolysis

Hydrolysis of esters and amides in the presence of serine proteases and thiol proteases has been widely studied. Papain is the best understood of the thiol protease family. The reaction occurs by the following mechanism (*4*):

$$E + S \underset{k_{-1}}{\overset{k_{+1}}{\rightleftharpoons}} E.S \xrightarrow{k_2} EA_C + P \xrightarrow{k_3} E + Acid \quad (1)$$

where ES and EA_C are respectively the non covalent Michaelis complex and the covalent acylenzyme complex and P is an alcohol or an amine.

Applying the steady state assumption to EA_C gives:

$$v = [E]_0 [S] \left[\frac{k_2 k_3/(k_2 + k_3)}{K_S k_3/(k_2 + k_3) + [S]} \right] \quad \text{with } K_S = \frac{[E][S]}{[ES]} \quad (2)$$

This is a Michaelis-Menten equation:

$$v = k_{cat} [E_0][S] / (K_M + [S])$$

$$\text{where} \qquad K_M = K_S \ \frac{k_3}{k_2 + k_3} \qquad\qquad (3)$$

$$k_{cat} = \frac{k_2\,k_3}{k_2 + k_3} \qquad\qquad (4)$$

From (3) and (4) it can be deduced that:

$$\frac{k_{cat}}{K_M} = \frac{k_2}{K_S} \qquad\qquad (5)$$

$$\text{If} \ \ k_3 \ll k_2 \qquad\qquad\qquad\qquad (6)$$

$$K_M = K_S \ \frac{k_3}{k_2} \qquad (7) \qquad\qquad k_{cat} = k_3 \qquad (8)$$

The parameters k_{cat} and K_M can be determined by the Lineweaver-Burk plot.

The hydrolysis of esters will now be considered in detail. The molecular species involved in equation (1) are given in Scheme (1). The major roles in the catalysis are played by Cys-25 and His-159. A non-covalent Michaelis complex is formed by interaction between the ester carbonyl and the sulfhydryl anion. Nucleophilic attack by the anion and acid catalysis by the protonated His-159 gives the acylenzyme EA_C which then undergoes nucleophilic attack by a water molecule, the final products being acid and the initial enzyme.

The structure of the active site and its interactions with the substrate (Michaelis complex) will be comprehensively considered elsewhere (5). On the basis of a crystallographic study by Drenth et al. (3) and Stubbs et al. (6) of the complexes of α-chloroketone and stefin β with papain, a model for the interactions in the active site was proposed (7): to each bonding site names P_1, P'_1, P_2, P'_2... in the substrate corresponds a bonding site S_1, S'_1, S_2, S'_2...of the enzyme. Cyst-25 (S_1) attacks at P_1. The prime sites usually correspond to the part of the molecule giving the alcohol.

The validity of equations (5) (6) and (8) was proven by data from Berezin et al (8), reported by Fersht (4). Esters of the type R'X-R where X represents amino acids (gly, but, norval, val, norleu, phe or tyr), R' is the acyl group of the amide (acetyl or benzoyl) and R is the alcohol residue of the ester group (methoxy, ethoxy, or isopropyloxy) have been hydrolyzed in the presence of α-chymotrypsine. Equation (5) was verified in all cases. Equation (8) is not verified if R' is a benzoyl group and inequality (6) does not apply if R' or R are respectively benzoyl or isopropyloxy groups. Both are verified in all other cases.

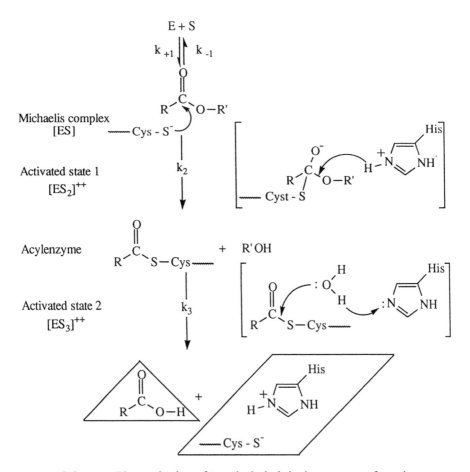

Scheme 1. The mechanism of ester hydrolysis in the presence of papain.

Experimental

Synthesis and hydrolysis. The synthesis of the dimethylester of N-succinyl-L or D-phenylalanine (1) and (2) and of the other derivatives will be described elsewhere (5). The melting point of products (1) and (2) is 76-79°C. DL-phenylalanine methyl ester (3) and phenylalanine (4) are commercial products from Aldrich. N-succinylphenylalanine (5), L-phenylalanine methylsuccinate (6), the methyl ester of DL-phenylalanine-N-succinic acid CH_3O-Phe-CO-$(CH_2)_2$-CO-OH (7) were also synthesized to identify the site of scission in products (1) and (2) by analysis of the reaction medium after different times of incubation, using thin layer chromatography-densitometry for the detection and the quantitative analysis of the hydrolysis products. The absence of products (3) and (4) shows that rupture of the amide bond does not occur. For the L-dimethylester, traces of the diacid in (6) indicates that breaking of the phenylalanine ester was at least ten times more rapid than breaking of the succinyl ester. In the D-dimethylester, (7) was the lone product identified.

These results allowed the kinetics of the hydrolysis reaction to be followed by pHmetry. The reaction medium was a fresh solution composed of activators (2mM EDTA, 5mM cysteine), 300 mM NaCl, 0.4 mM sodium azide, enzyme (4.8 x 10^{-4} mM for the L-diester and 3.6 x 10^{-2} mM for the D-diester), and the soluble substrate. The substrate concentration was between 1 x 10^{-2} and 1.6 x 10^{-3} M for the L-diester and between 1 x 10^{-2} and 2.6 x 10^{-3} for the D-diester. The final concentration of enzyme was 4.8 x 10^{-7} M (0.01 mg/ml) for the L-diester and 3.6 x 10^{-5} M (0.75 mg/ml) for the D-diester. Its specific esterase activity determined using benzoyl-L-arginine ethylester as substrate (15)was 34 µmol/min.mg. The pH evolution was followed with a microprocessor CG840 Schott pHmeter from 6.2 to 6.0. The pH was manually readjusted to the initial value by addition of a measured quantity of 5 x 10^{-2} M NaOH. Experiments carried out in the same conditions with the exception of the enzyme solution which was replaced by water, showed that chemical degradation does not occur in the experimental conditions used in the present work.

Computer simulation. Computer simulation of enzyme substrate interactions have been performed using two softwares: conformational analysis by the tree structure method (9) and Hyperchem using the AMBER 3.0.a force field.

The first method is a semi-empirical conformational analysis method developed by Brasseur for the study of interactions between amphiphile molecules with membrane lipids and was applied to various systems including phospholipids, pharmacological agents and peptides (10). The molecule is defined using standard values of interatomic distances, valence and torsion angles and net charges assigned to each atom. CH_3, CH_2 and CH groups are considered as atomic entities with the exception of asymetric carbons. Total conformational energy is considered to be the sum of Van der Waals and coulombic interactions, torsion potentials and variations of transfert energies from a hydrophobic to a hydrophilic environment. This is calculated for a large number of conformations in a systematic analysis called "*tree structure*".

Calculations in Hyperchem are based on the potential energy surface of a system as well in the geometrical optimization algorithms as in those of molecular dynamics. This surface is calculated using molecular mechanics. AMBER force field first developed by Weiner et al. (*11*) is given by the following expression:

$$E = \sum_{bonds} K_r (r - r_{eq})^2 + \sum_{angles} K_\theta (\theta - \theta_{eq})^2 +$$

$$\sum_{dihedrals} \frac{V_n}{2} [1 + \cos(n\varphi - \varphi_0)] + \sum_{ij} \left[\frac{A_{ij}}{R_{ij}^{12}} - \frac{B_{ij}}{R_{ij}^{6}} \right]$$

$$+ \sum_{ij} \frac{q_i\, q_j}{\varepsilon\, R_{ij}} + \sum_{ij} \left[\frac{C_{ij}}{R_{ij}^{12}} - \frac{D_{ij}}{R_{ij}^{10}} \right] \qquad (9)$$

where the successive terms represent bond stretching, bending, torsional, Van der Waals, electrostatic and H-bonding energies. To the force field is associated a library which gives the value of the various parameters given in equation 9 for each type of atoms. H-bonds are easily visualized on the screen by dashed lines when the distance separating donor and acceptor atoms are lower than 3.2Å and when the larger dihedral angle formed between these two atoms and a third one linked to one of them (C = O ... H or O ... H - N) is larger than 150°.

The purpose of the algorithm used in geometrical optimization is to annulate the derivative of the potential energy with respect to the cartesian coordinates of the atoms in order to find the conformation corresponding to the minimal potential energy of the system. Successive energy minimizations were performed. Each optimization was stopped when the gradient is lower than 0.1 Kcal/mol/A. Entropy effects are not taken into account by the algorithm. Solvation effects were introduced by using a distance-dependent dielectric constant for the electrostatic energies.

The purpose of molecular dynamics was to simulate the dynamic behaviour of the molecules by incorporating the thermal movement of the atoms, allowing them to pass potential energy barriers using Newton's movement equations. The algorithm calculates the next position and rate of atoms after a time increement δt using the preceding value of these parameters. Conformational changes can be studied by operating successive time increements δt corresponding to a total time of a few ps. In the present case, molecular dynamics was performed after the second step of geometrical optimization. The total time was 10 ps. The conformation of the molecules but also the value of the various components of the energy of the system were recorded every 0.5 ps a total number of 21 values being thus obtained. For practical reasons of computer time, molecular dynamics was performed only on amino-acids of the enzyme active site and on atoms of the substrate.

Results

Kinetics of Hydrolysis. As demonstrated in the experimental part of this work, the main position of scission for the L- and D- dimethylester of N-succinylphenylalanine are: the ester bond at the phenylalanine site for the derivative (L-Phe) and at the succinyl site for the D-derivative (D succ)

$$CH_3O \dashv CO-CH-NH-CO-(CH_2)_2-CO-OCH_3 \qquad CH_3O-CO-CH-NH-CO-(CH_2)_2-CO \vdash OCH_3$$
$$\qquad\quad CH_2 \qquad\qquad\qquad\qquad\qquad\qquad\qquad\qquad CH_2$$
$$\qquad\quad Ph \qquad\qquad\qquad\qquad\qquad\qquad\qquad\qquad\quad Ph$$
$$\qquad\quad (L\text{-}Phe) \qquad\qquad\qquad\qquad\qquad\qquad (D\text{-}Succ)$$

The product resulting from the scission of the L-diester at the succinyl bond (L-succ) is also observed in small quantity at long time indicating that the reactivity of this bond is much lower than the other by a factor of hundred. Scission of the phenylalanine methylester (D-Phe) was not observed for the D-diester.

Initial rates ($v_0/[E_0]$) were calculated using four initial substrate concentrations. The inverse initial rates vary linearly with 1/[S] as required by Michaelis-Menten kinetics. The values of K_M, k_{cat} and K_M/k_{cat} calculated from the regression lines are given in Table I. For these concentrations, the condition $[S] \ll K_M$ is verified in all cases.

Table I shows that the k_{cat} value for hydrolysis of the succinic methylester (2) derived from D-phenylalanine is about hundred times lower than k_{cat} obtained from hydrolysis of the phenylalanine methylester (1) derived from L-phenylalanine. This implies a factor of hundred in the specificity constant k_{cat}/K_M, K_M having neighbouring values for (1) and (2).

These values of k_{cat}, K_M and k_{cat}/K_M can be compared with other values of the literature (7) for L-derivatives also given in Table I. A first examination of examples 8 to 14 shows that:
1. the nature of the amide substituent (CH_3-O-CO- or CH_3-CO) has a limited effect on k_{cat} and K_M (compare 8 and 9).
2. the presence of glycine has a very strong influence, scission at the glycine ester group (next but one to phenylalanine) being strongly favoured (compare 9 and 10).

3. the replacement of the peptide bond CO-NH by ester CO-O adjacent to phenylalanine has a very strong importance on the rate (compare 9 and 11).
4. the presence of phenylalanine strongly increases the rate (compare 9 and 13); (compare 9 and 14).

The differences in rate between examples 8 to 14 are mainly due to differences in K_M rather than to differences in k_{cat}. The broad differences in reactivity of these derivatives have been assigned by Berti et al. (7) to the number and nature of hydrogen bonds between the substrate and the enzyme. Formation of these bonds on passing from E + S to ES or to the activated state leads to specific increases in enthalpy and decreases in entropy.

If we now compare the reactivity of products (1) and (2) with the other compounds in Table I, we see that k_{cat} for (1) is of the same order of magnitude as the others while its K_M is much larger. Comparison of structures (1) and (12) shows that in both cases the ester bond which was be broken was adjacent to phenylalanine. Both molecules contain one amide or one urethane group, and two ester groups. The k_{cat}/K_M constants are of a same order of magnitude, on the low rate side of table I and much lower than that of products (8) and (9). As for compound (2), it is the least reactive of those reported in table I.

As stated above, an increase in rate (k_{cat}/K_M) can be due to an increase of k_{cat}, a decrease of K_M or both simultaneously. The third case is illustrated by the date of Berezin et al cited previously (8). As the size of the hydrophobic group in the amino acid is increased, k_{cat}/K_M increases over a range of 10^6, the effect being distributed between lowering K_M and increasing k_{cat}.

Computer Simulation. The initial structure of the substrates was obtained by the tree structure method (9) followed by two successive geometrical optimizations. The obtained structure are given in fig. 1 and 2. The potential energies of the L- and D-diesters are respectively - 4.66 and -6.05 kcal/mol.

The initial structure of papain was determined by Priestle et al (12). The resolution of the structure is 2Å and the root mean square deviation from ideality is 0.007Å for the bond lengths and 1.9° for the bond angles. Modifications have been brought to the structure: the blocking agent has been withdrawn and the sulfur atom of Cys-25 modified into sulfhydryl anion; His-159 has been transformed in order obtain the imidazolium cation. The charges on the enzyme have been calculated using AMBER force field.

The enzyme-substrate Michaelis complexes have been simulated for four situations, each of them favouring the hydrolysis of one of the four ester bonds. They are: hydrolysis of the phenylalanine ester for L- and D-diesters (L-Phe and D-Phe) and hydrolysis of the succinyl ester for the L- and D diesters (L-succ and D-succ). The various complexes have been built by using the following rules, the first one being the most important:

• Cys-25 and the bond to be hydrolyzed are situated near each other (3Å)
• Hydrogen bonds P2-S2 and hydrophobic interactions are favoured..

TABLE I: Kinetic constants for hydrolysis of phenylalanine derivatives by papain

Substrate	k_{cat} (s^{-1})	K_M mM	k_{cat}/K_M $(M^{-1}s^{-1})$	Ref
(1)CH$_3$O-Phe-CO-(CH$_2$)$_2$-CO-OCH$_3$(L)[a]	7.3	100	73	d
(2)CH$_3$O-Phe-CO-(CH$_2$)$_2$-CO-OCH$_3$(D)[a]	0.07	120	0.65	d
(8)CH$_3$O-gly-Phe-CO-CH$_3$[ab]	9.4	0.078	120.000	e
(9)CH$_3$O-gly-Phe-Moc[abc]	6.3	0.16	39.000	e
(10)CH$_3$O-Phe-Moc[ac]	-	-	7.12[f]	e
(11)CH$_3$O-CO-CH$_2$O-Phe-Moc[ac]	-	-	430	e
(12)CH$_3$O-CO-CH$_2$-O-Phe-Moc[ac]	-	-	52	e
(13)CH$_3$O-gly-gly-Moc[bc]	1.8	45	40	e
(14)CH$_3$O-gly-N-benzoyl[b]	2.8	20	140	e

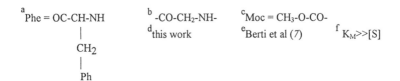

[a]Phe = OC-CH-NH   [b] -CO-CH$_2$-NH-   [c]Moc = CH$_3$-O-CO-

         | [d]this work   [e]Berti et al (7)   [f] K_M>>[S]

        CH$_2$

         |

         Ph

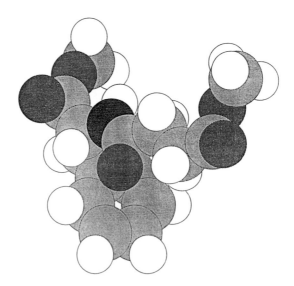

Figure 1. Three-dimensional structure of the D-dimethylester of N-succinyl-phenylalanine

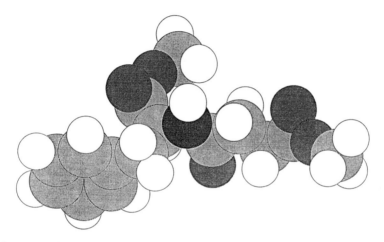

Figure 2. Three-dimensional structure of the L-dimethylester of N-succinyl-phenylalanine

The geometrical optimization has been realized in two steps. In the first one, the substrate and the following amino-acids situated in the active site have been considered (18-29, 64-69, 131-137, 157-180, 175-177, 204-207), the other amino-acids remaining unchanged. In the second step, the whole protein and the substrate are involved in the geometrical optimization. After the second step, molecular dynamics has been performed on the amino-acids of the active site, assuming that they are responsible for the evolution of the substrate conformation. The system is considered to be stationary if the fluctuation interval of the potential energy with respect to the total value is lower than 5%. The hydrophobic interactions are evaluated by considering the proximity of the phenyl group of phenylalanine and of the side chains in the hydrophobic pocket. The formation of H-bonds, as calculated by Hyperchem, can be visualized on the screen and evaluated in statistical terms. We have 21 instantaneous conformations in each dynamics (every 0.5 ps during 10 ps). The ratio of the number of time one given H-bond could be observed with respect to the total number of instantaneous conformation gives the frequency of formation of this bond. The results are given in Table II.

Conformation of the enzyme-substrate Michaelis complex. The structure obtained for the enzyme-L-diester complex in the phenylalanine approach is given in fig. 3. The substrate and Cys-25 sulfur, nitrogen, carbon and oxygen are represented by spheres. The remaining enzyme is given as a thread corresponding to the polypeptide chain. It can be observed that the active site is situated near the surface of the enzyme. It has the form of a cleft the surface of which is covered by amino acids able to form H-bonds or hydrophobic interactions with the substrate. The carbonyl group to be hydrolyzed lies near the sulfhydryl anion of Cys-25. In order to better visualize the intermolecular interactions at the active site, the structures obtained after the third geometrical optimization for two of the four complexes corresponding to the L-diester in the phenylalanine and in the succinyl approach are given in figures 4 and 5. The H-bonds that are apparent in the given orientation of the complex are indicated by dashed lines. The types of H-bonds and their frequency of formation are given in Table II. It can be observed that the phenylalanine part of the molecule lies in the active site in the phenylalanine approach confirming figure 3 while it is rejected outside the slit for the succinic approach. Qualitatively, the same type of images (aromatic ring in the clift for the phenylalanine approach and out of it for the succinyl approach) have been obtained for the D-diester and are not given here. Hydrophobic interactions are thus possible for both derivatives in the former approach. Stereospecificity for the L-derivative in the phenylalanine approach is thus not due to the absence of hydrophobic interactions with the D-derivative as stated by other authors (*13,14,4*).

H-bond formation can now be considered to justify this stereospecificity in the phenylalanine approach. If the L- and D-derivatives are compared in Table II, it appears that the type, frequency of formation and proximity of donor and acceptor in H-bonds are very different for both derivatives. The number and frequency of H-bonds in the D-derivative are lower than those aformed with the L-derivative. It can

Table II. Nature and frequency of H-bonds (molecular dynamics)

Site	Bond		L-Phe	D-Phe	L-succ	D-succ
Gly66	ester C=O... HN (gly66)	l_1	19	14	43	57
	amide NH...O = C (gly66)	l_2	50	0	0	0
	ester CO-O...HN (gly66)	l_7	10	0	52	33
	$\quad\mid$					
	$\quad$CH$_3$					
	amide C=O...HN (gly66)	l_8	14	0	0	0
	amide NH HN (gly66)	l_9	0	19	0	0
Asp.158[a]	amide NH... O = C (Asp158)	l_6	0	19	0	0
	(Asp158 SG) C= O...HN (His159 MC)	l_{10}	33	10	33	10
	(gly19) C = O...HN (His159 SG)	l_{11}	57	29	38	48

H-bond: d < 3.2Å; angle > 150°
[a]MC: main chain; SG: side group

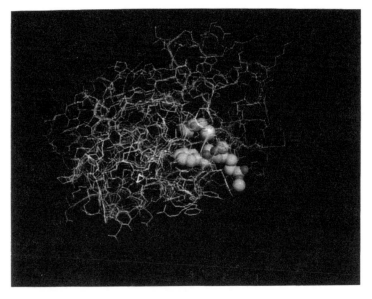

Figure 3. Three-dimensional structure of the L-dimethylester in papain before molecular dynamics (phenylalanine approach).

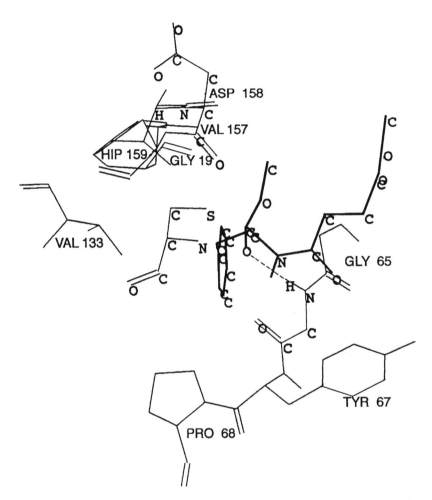

Figure 4. Three-dimensional structure of the L-dimethylester in the active site of papain after molecular dynamics (phenylalanine approach).

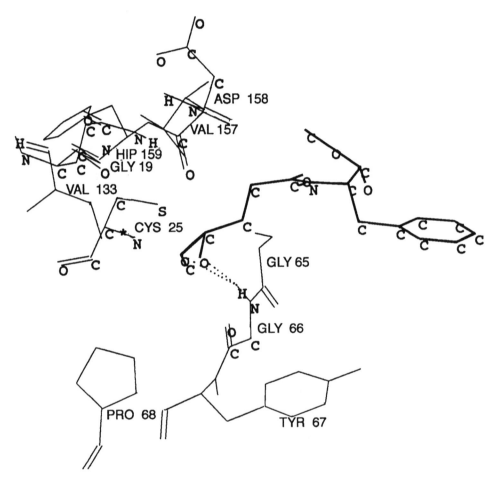

Figure 5. Three-dimensional structure of the L-dimethylester in the active site
of papain after molecular dynamics (succinyl approach).

thus be proposed that H-bonding is globally responsible for stereospecificity. If the succinyl approach is now considered, it appears that H-bonds with the amide bond of the diester cannot form because the NH-CO group lies out of the active site.

New intramolecular H-bonds between amino-acids of the enzyme are formed in the presence of the substrate (l_{10} and l_{11}). They show less variability as a function of the type of approach than the intermolecular bonds.

Potential energy of the complexes. The potential energies of the enzyme-substrate complexes resulting from the third geometrical optimization which follows molecular dynamics are given in table III for the substrate in the whole enzyme. Similar calculations have been performed for interactions between the substrate and only the active site of the enzyme. Energy has of course a much larger negative value in the first case than in the last one but the sequence is the same. The main contribution to potential energy is the electrostatic term which contains a contribution from H-bonds. Indeed, the main component of the H-bonds is an electrostatic interaction between the dipole of the covalent bond to the hydrogen atom, in which the hydrogen atom has a partial positive charge, and the dipole of the other covalent bond where a partial negative charge lies on the donor electronegative atom.

The more reactive (L-phe) approach corresponds to the less stable complex and the less reactive (D-Phe) to the more stable complex. The L-Phe approach was already characterized by the highest potential energy after the first geometrical optimization and kept it through the successive calculation steps.

Qualitative Gibbs free energy diagram is given in figure 6 for the various steps of the hydrolysis of L-derivative in the phenylalanine approach, equations 5, 6 and 7 being taken into account. Let us remember that $k_{cat} = k_3$, $K_M = K_S k_3/k_2$ and $k_{cat}/K_M = k_2/K_S$ have been measured. Gibbs activation free energies ΔG^{++} corresponding to the rate constants can be calculated by application of the general relation $k = (kT/h) e^{-\Delta G^{++}/kT}$. Gibbs free energy ΔG^{++} corresponding to equilibrium constants are given by $K = e^{-\Delta G/RT}$. The potential free energy contribution to K_S has been calculated. Indeed, the previously calculated potential energies (Table III) are the energy contribution to the total free energy of ES. If the entropy difference between E + S and ES is assumed to be the same for the various approaches, it can be assumed the free energy ΔG_S for the dissociation of ES increases with the corresponding difference in potential energy. These values of $[E_{pot}(E) + E_{pot}(S)] - E_{pot}(ES) = \Delta E_{pot}$ are given in table IV.

The rate and equilibrium parameters k_2 and K_S for the L-derivative in the phenylalanine approach can tentatively be compared with those for the D-derivative in the succinyl approach. If the parameters characterizing the L- and D-derivatives are respectively k_2 and k'_2, K_S and K'_S, experimental results of table I give, if equation 5 is taken into account:

Table III. Potential energy (total enzyme + substrate) obtained by computer simulation (kcal/mol)

Substrate (approach)	L-Phe	L-succ	D-succ	D-Phe
Contribution				
Bond stretching	41.1	42.4	42.5	40.7
Bond bending	311	312	312	305
Torsional energy	261	272	272	262
V.d.W. energy	-953	-939	-947	-947
H. Bonding	-88.6	-92.6	-92.5	-88.2
Electrostatic energy	-3202	-3282	-3294	-3298
Total	-3631	-3687	-3737	-3726

Table IV. Potential energy difference for the dissociation of ES

Approach	[Epot(E) + Epot(S)]-Epot(ES)	Ester bond hydrolyzed
L-Phe	-59.2	Phenylalanine
L-Suc	+0.74	(succinyl)
D-Phe	+38.2	Phenylalanine
D-Suc	+18.0	succinyl

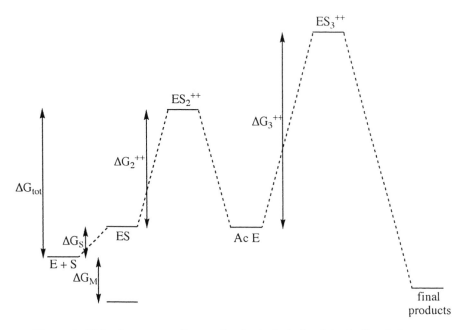

Figure 6. Gibbs free energy diagram for the hydrolysis of the L-dimethylester in the presence of papain (phenylalanine approach).

$$\frac{k_2}{K_S} > \frac{k'_2}{K'_S} \tag{10}$$

The data obtained by computer simulation and given in table IV indicate that $K_S > K'_S$ since $\Delta E'_{pot} > \Delta E_{pot}$ and entropy differences are assumed to be equal for both components. The consequence of this is that k_2 has to be much larger than k'_2 in order to verify inequality equation 10. It follows that ΔG_2^{++} is smaller than $\Delta G_2^{++'}$. The physical meaning of this could be the following. The substrate has to be favourably oriented in the active site to undergo the nucleophilic attack by the sulfhydryl anion; a too high stability of the non covalent complex (high ΔG_S and low K_S) results in a decrease of the rate of hydrolysis (high ΔG_2^{++} and low k_2).

The rate of enzymatic reactions has been discussed by Fersht (4) in terms of complementarity between enzyme and substrate in the Michaelis complex and in the transition state. Increased complementarity means increased binding energy and lower potential energy of the entity either ES or ES^{++}. As discussed previously in this paper, an increase of k_{cat}/K_M which is equal to k_2/K_S can result from an increase of k_2, a decrease of K_S, an increase of k_2 combined with a decrease of K_S or an increase of K_S combined with an important increase of k_2. These various situations together with the associated changes in complementarity are summarized in figure 7. The last case, (increase of K_S combined with an important increase of k_2) corresponds to the results of the present work. It has been demonstrated by Fersht (4) that this case is the most favourable one to obtain maximization of rate in the most general case where the rate in given by:

$$v = \frac{k_{cat}}{K_M} \ [E][S]$$

The value of k_{cat}/K_M is first maximized by having complementarity between enzyme and substrate at the level of the transition state. Then [E] is maximized by having a large value of K_M.

To our knowledge the present work is the first attempt to simulate Michaelis complexes and to calculate their potential energy. Computer simulation of other systems such as those given in (7) and (8) would contribute to test the validity of the theoretical part of this work.

Conclusions

L- and D-dimethylesters of N-succinylphenylalanine are interesting examples of the relative reactivity of the various esters and amide bonds to be hydrolyzed in the presence of papain. In the conditions used in the present work, the amide group is not hydrolyzed and is thus less reactive than the ester groups as reported previously (4) for other derivatives. The L-phenylalanine ester is about ten times more reactive

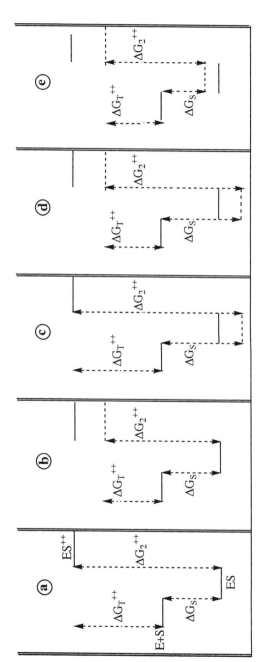

—— initial level

----- modified level

a : initial system

b : complementary to ES^{++} increases -> K$_S$ is constant, k$_2$ and $\frac{k_2}{K_S}$ increase

c: complementary to ES increases -> K$_S$ and k$_2$ decrease; $\frac{k_2}{K_S}$ is constant

d: complementary to ES^{++} and ES increases -> K$_S$ decreases, k$_2$ and $\frac{k_2}{K_S}$ increase

e: complementary to ES decreases and to ES^{++} increases-> K$_S$, k$_2$ and $\frac{k_2}{K_S}$ increase

Figure 7. Effect of complementarity to ES and ES^{++} on K$_S$, k$_2$ and k$_2$/K$_S$

than the D- and L- succinylesters. Hydrolysis of the phenylalanine ester of the D-derivative is not observed. This stereospecificity for the phenylalanine esters has been assigned to differences in H-bonding of the L- and D-diesters in the active site of papain on the basis of computer simulation. The computed conformation of the Michaelis complex shows that in the succinyl approach, the aromatic ring and the amide group of both L- and D-derivatives lie out of the active site, thus altering substrate-enzyme interactions. The kinetic reactivity of the various ester sites decreases with decreasing computed potential energy of the non covalent Michaelis complex. From a kinetic point of view, the higher reactivity of the L-phenylalanine ester when compared to the D-succinyl ester is due to an increase of K_S combined with an important increase of k_2. This corresponds to a decreased complementarity in the Michaelis complex and increased complementarity in the activated state.

Acknowledgments. The authors are very grateful to Dr. A. Crucq, Senior Research Associate at the Fonds National de la Recherche Scientifique, for very stimulating discussions.

Literature Cited.
(*1*)   Gilkes, N.K. Jervis, E., Henrissat, B. Tekant, B., Miller, R.C., Jr., Warren, R.A.J. and Kilburn, D.G. , *J. Biol. Chem.* **267**, 6743 (1992).
(*2*)   David, C. and Fornasier, R.. *Macromolecules* **19**, 552 (1986).
(*3*)   Drenth, J., Jansonius, J.N., Koekoek, R. and Wolthers, B.G. *Adv. Protein Chem.* **25**, 79 (1971).
(*4*)   Fersht, A.F. In *Enzyme Structure and Mechanism.* W.H. Freeman and Company, 2d Edition, New York (1984).
(*5*)   Lefebvre, F., David, C. et al. to be published.
(*6*)   Stubbs, M.T., Laber, B., Bode, W., Huber, R., Jerabe, R. Lemarcie, B. and Turk, V., *EMBO J.* **9**, 1939 (1990).
(*7*)   Berti, P.J., Faerman, C. and Storer, A.C. *Biochemisry*, **30**, 1394 (1991).
(*8*)   Berezin, I.V., Kazanskaya N.F.and Klyosov, A.A. *FEBS Lett.* **15**, 121 (1971).
(*9*)   Brasseur, R. and Deleers, M. *Proc. Natl. Sci.* (USA) **81**, 3370 (1984).
(*10*) Brasseur, R.. In *Molecular Description of Biological Membrane Components by Computer Aided Conformational Analysis.* Vols. 1-2 CRC Press Inc. 2000 Corporate Blvd. N.W, Boca Raton,Florida 33431.
(*11*) Weiner, S.J. Kollman, P.A., Case, D.A., Singh, U.C., Ghio, C., Alagona, G., Profeta, S., Weiner, P. *J. Am. Soc.*, **106**, 765, (1984).
(*12*) Priestle, J.P., Ford, G.C., Glor, M., Mehler, E.L., Sonet, J.D.G., Thaller, N. and Jansonius, J.N. *Acta Cryst.*, Section A, **40** (1984).
(*13*) Berger, A. and Schechter, I., *Philos. Trans. R. Soc. London*, **B259**, 249 (1970).
(*14*) Lowe, G. and Yuthavang, Y., *Biochem. J.* **124**, 107 (1971).
(*15*) Arnon, R. In *Methods of Enzymology* **19**, 226, Academic Press, New York (1970).

POLYMER MODIFICATIONS

Chapter 9

Regioselective Enzymatic Transesterification of Polysaccharides in Organic Solvents

Ferdinando F. Bruno[1], Jonathan S. Dordick[2], David L. Kaplan[3], and Joseph A. Akkara[1,4]

[1]Biotechnology Division, U.S. Army Soldier Systems Command, Natick Research, Development, and Engineering Center, Natick, MA 01760–5020
[2]Department of Chemical and Biochemical Engineering and Center for Biocatalysis and Bioprocessing, University of Iowa, Iowa City, IA 52242
[3]Biotechnology Center, Department of Chemical Engineering, Tufts University, 4 Colby Street, Medford, MA 02155

Enzymes catalyze a large number of reactions in non-aqueous media that are not possible in aqueous solution. One of these reactions is the enzyme catalyzed acylation of alcohols. Recently it was observed that enzymes can be extracted from aqueous solutions into organic solvents by using very low surfactant concentrations via ion-pairing between the surfactant and the protein. In the present work we demonstrate the site of selective transesterification of polysaccharides with fatty acids in the presence of ion-paired subtilisin carlsberg. The polysaccharides and corresponding derivatives used include amylose, cyclodextrins, and hydroxyethyl cellulose. Characterization data including FTIR and TGA is reported.

Earlier studies have indicated that subtilisin, a protease, is a useful catalyst in organic solvents to carry out synthetic organic chemistry (1). In organic solvent reaction media this enzyme catalyzes the formation of esters, while in the presence of water, the enzyme hydrolyzes esters (2). Moreover, substrates, such as monosaccharides, used in esterification reactions were soluble in organic solvents, permitting interaction between the soluble substrate and the insoluble enzyme (3). The formation of ion-pairs between an enzyme like subtilisin Carlsberg (from *Bacillus licheniformis*) and a surfactant, was recently studied (4). This catalyst system in organic solvent maintains its activity for the acylation of polysaccharides and oligosaccharides in organic solvents (5). Polysaccharides are of interest because of their biodegradability and biocompatibility properties as well as their wide ranging applications as coatings, finishes, membranes and fibers. Enzyme modified polysaccharides can be useful as biodegradable emulsifiers, compatibilizers, and detergents as well as for surface modification of preformed polysaccharides-based materials (6).

[4]Corresponding author.

167

In this work, we demonstrate that amylose, β-cyclodextrin and hydroxy ethyl cellulose (HEC) (three organic solvent-insoluble polysaccharides consisting of α or β-1,4 linked glucose moieties), when cast as thin films or suspended as cryogenically milled powder, can be transesterified in organic solvents by ion-paired subtilisin protease. Moreover, the results indicated that amylose was regioselectively acylated. These are the first attempts at modifying solvent-insoluble polymers catalyzed by enzymes in organic solvent reaction media.

Materials and Methods.

Subtilisin Carlsberg protease enzyme (Amano Enzyme Co., Troy, VA, 1.1 mg/mL) was dissolved in HEPES buffer (8.5 mM, pH 7.8) containing 6 mM KCl. The aqueous solution was mixed with an equal volume of isooctane containing 2 mM dioctyl sulfosuccinate sodium salt (AOT) and the two-phase solution was stirred at 250 rpm at 25°C. After 30 minutes of stirring, the phases were allowed to settle and the organic phase was removed (5). The protein and water content of the solution were determined by absorbance at 280 nm and Karl-Fischer titration, respectively. Approximately 1.0 mg/mL of enzyme was present in the isooctane solution with a water content of <0.01% (v/v). Amylose was dissolved in water and dried onto ZnSe slides as a thin film approximately 1 μm in depth. The reactions were performed with 1.0 mg/mL ion-paired subtilisin protease in isooctane containing 60 mM n-capric acid vinyl ester (C10VE, TCI America, Portland, OR) in a beaker containing the ZnSe slide without shaking for 48 h at 37°C. Similar procedures were used for the β-cyclodextrin and for HEC. The solid amylose film was removed from the beaker and washed with fresh isooctane to extract unreacted vinyl ester. Reactions were also conducted with the polysaccharide in powder form, using similar procedures. The polymer powders had a particle size of less than 100 μm with a surface area to weight ratio of 546 cm^2/g.

Results and Discussion

As represented in Fig. 1, the Fourier Transformer Infrared (FTIR) spectrum indicated the formation of a derivatized β-cyclodextrin by the enzyme reaction based on large absorptions at 2920 and 2850 cm^{-1}, corresponding to the methylene stretch of an alkyl chain in the final product. In addition, the presence of a peak in the region 1693-1730 cm^{-1}, indicates a C=O group. Spectra of unmodified β-cyclodextrin or β-cyclodextrin treated with the vinyl ester in the absence of enzyme (Figure 1) do not exhibit these absorptions peaks. Additionally, the lack of absorbance peaks at 871 and 951 cm^{-1} in the FTIR spectrum indicate the absence of vinyl groups in the modified β-cyclodextrin. Therefore, there was no non-selective adsorption of the vinyl esters to the β-cyclodextrin. Similar results were found for amylose (5) and for HEC.

In addition to the β-cyclodextrin film, cryogenically milled β-cyclodextrin powder was also studied as an isooctane insoluble substrate. The FTIR spectrum of the β-cyclodextrin powder, after the enzymatic reaction, also exhibited evidence of

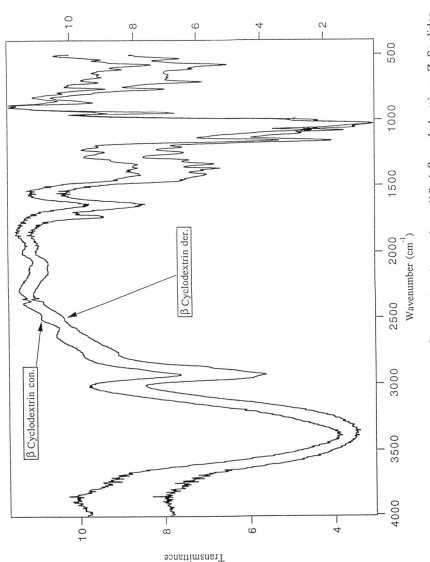

Figure 1. FTIR spectra of unmodified β–cyclodextrin, and modified β–cyclodextrin on ZnSe slides.

acylation analogous to that found in Fig. 1. When subtilisin protease powder, in the absence of AOT (during the extraction procedure), was used as a catalyst, no acylation reaction was observed. Specifically, in the presence of 60 mM vinyl caprate and 10 mg/ml of subtilisin Carlsberg protease powder (without AOT and no ion pairing), no β-cyclodextrin derivatization was apparent when β-cyclodextrin was used either in powder or thin film forms. Consequently, solubilization of the enzyme in an organic solvent in the presence of AOT is important for acylation of the insoluble polymer.

The degree of substitution catalyzed by the enzyme reactions in the powdered β-cyclodextrin and amylose was assessed semiquantitatively by thermogravimetric analysis (TGA) and for amylose also by Electron Spectroscopy Chemical Analysis (ESCA). The TGA profile for native β-cyclodextrin and enzymatically acylated β-cyclodextrin powders are shown in Figure 2. The major difference in the thermograms between the derivatized and unmodified β-cyclodextrin (prior to backbone thermal degradation) was the weight loss in the former at 280°C. This weight loss is characteristic for alkyl chain degradation (7) and was not present in the native β-cyclodextrin. Quantitation of the weight loss of the modified β–cyclodextrin as compared to the unmodified β-cyclodextrin indicates that about 0.18 acyl groups are associated per glucose moiety. Previously, similar results were found for the enzyme-modified amylose acylation by enzyme (6). Chemically and non regioselectively acylated β-cyclodextrin (using acyl chlorides), with higher degree of substitution, shows similar TGA profiles as those reported for the enzymatically-treated polymers (Fig. 2).

These results indicated acylation of surface accessible hydroxyl sites on the polymer chains, a small fraction of the total amylose present in the reaction film. Therefore, ESCA analysis was used to characterize the top 100 Å of the amylose film. This analysis indicated that the acylated surface had a degree of substitution of 0.9±0.1 acyl groups per glucose moiety, based on the C:O ratio. It has been reported that subtilisin protease selectively acylated primary hydroxyl groups in reaction with sugar carried out in organic solvents (8). Therefore, based on the ESCA analysis, it was possible to speculate that regioselective acylation of the primary hydroxyl groups on the polysaccharides also occurred in our procedure. Consequently, ^{1}H-NMR was used to determine the position of enzymic acylation of amylose. In comparison to underivatized amylose, the only significant shift observed was in the 6-hydroxyl proton of the modified amylose. ^{1}H-NMR of the enzymatically modified amylose powder and β–cyclodextrin show peaks at 0.8 and 1.2 ppm representing CH_3 and CH_2 protons, respectively, confirming the presence of a straight-chain moiety on the derivatized amylose and β-cyclodextrin. Such groups do not exist in the unmodified polymers.

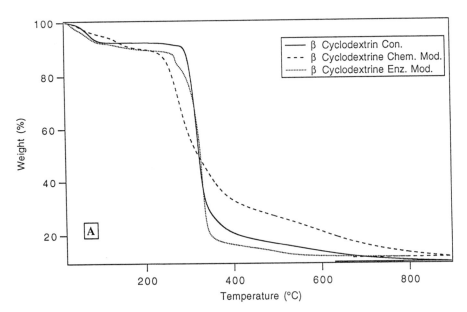

Figure 2. Thermogravimetric analysis of (A) unmodified, enzymatically modified and chemically modified β–cyclodextrin and (B) derivatives (Δ weight/temp.) of unmodified, enzymatically modified and chemically modified β–cyclodextrin.

continued on next page

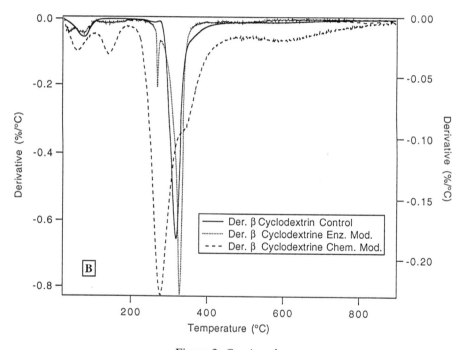

Figure 2. *Continued*

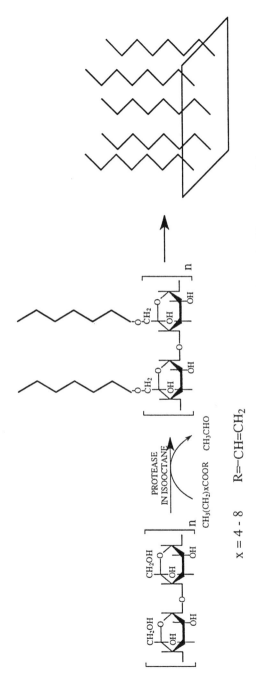

Figure 3. Proposed structure of modified polysaccharides prepared by subtilisin Carlsberg protease catalysis.

Enzymatic specificity of acylation for amylose was demonstrated by ^{1}H-NMR
(*DMSO-d6*): (a) native amylose : ∂ 3.8 (2 H, 3 H, 4 H, 6 H, br), 4.3 (6 H, m), 4.5
(6 OH, m, area 0.348), 4.9 (1 H, ax, m), 5.2 (1 H, br, area 0.336), 5.5 (3 OH, br,
area 0.330), 5.6 (2 OH, br, area 0.330);

(b) derivatized amylose : ∂ 0.8 (CH$_3$, br), 1,2 (CH$_2$, br), 1.3 (CH$_2$, br), 2.2
(CH$_2$, br), 3.75 (2 H, 3 H, 4 H, 6 H, br), 4.4 (6 OH, m, area 0.467), 5.2 (1 H, br,
area 0.509), 5.4 (3 OH, br, area 0.495), 5.45 (2 OH, br, area 0.495). It should be
noted that the area ratio of the 6-OH proton to total protons in the native and
derivatized amylose is 0.26 and 0.23, respectively. This finding provided additional
evidence that acylation was confined to the 6-OH group in this films.

Integration of the alkyl chain protons of the amylose derivatives and the
β–cyclodextrin protons resulted in a calculated degree of substitution of 0.185 and
0.250, respectively. This is slightly lower than that predicted by TGA analysis of the
powdered amylose. Such a discrepancy may result from the relatively qualitative
nature of TGA analysis as compared to ^{1}H-NMR.

Figure 3 illustrates the proposed structure of the modified amylose. Subtilisin
protease, when surfactant ion paired, appears to be highly efficient in catalyzing the
acylation of nearly all available surface primary hydroxyl groups in amylose thin
films or fine powder. This may provide a new route to modify surface properties of
polysaccharide films or fiber to control hydrophobicity and reactivity.

References

1. Klibanov, A. M. *Trends Biochem. Sci.* **1989**, *14*, 141.
2. Klibanov, A. M. *Acc. Chem. Res.* **1990**, *23*, 114.
3. Dordick, J. S. *Microb. Technol.* **1989**, *11*, 194
4. Paradkar, V. M.; Dordick, J. *Jour. Am. Chem. Soc.* **1994**, *116*, 5009.
5. Mayer, J.M.; Kaplan, D.L. *Trends Polym. Sci.* **1994**, 2, 227.
6. Bruno, F. F.; Akkara, J. A.; Kaplan, D. L.; Gross, R.; Swift G.; Dordick, J.
S. *Macromolecules* **1995**, *28*, 8881.
7. Lukaszewski, G. M. *Lab. Practice* **1966**, *15*, 551
8. Riva, S.; Chopineau, J.; Kieboom, A. P. G.; Klibanov, A. M.*Jour. Am.
Chem. Soc.* **1988**, *110*, 5849.

Chapter 10

Biocatalytic Modification of Alginate

Donal F. Day, R. D. Ashby, and J. W. Lee

Audubon Sugar Institute, Louisiana State University, Baton Rouge, LA 70803

Alginate is a major structural polysaccharide in many seaweed and is found as a capsular material secreted by some bacteria. The bacterial form has a higher molecular weight and has the mannuronic acid residues acetylated at the 2-0-position. It has been possible to use whole bacterial cells as biocatalysts and selectively acetylate the mannuronic acid residues of the seaweed polymer. This produces a polysaccharide with properties intermediate between the bacterial and seaweed polysaccharide.

Alginate is one of the few polysaccharides that can be obtained from both eukaryotes and prokaryotes. It is a structural polymer in numerous species of the brown seaweed, particularly members of the genera *Ascophyllum*, *Ecklonia*, *Fusarium*, *Laminaria*, and *Macrocystis*; constituting between 14 and 40% of the dry solids of these algae. The bacteria *Azotobacter vinlandii* (*1*) and several species of *Pseudomonas* (*2, 3*), including *Pseudomonas aeruginosa* (*4,5*) and *Pseudomonas syringae* (*6*), produce an alginate like polymer as extracellular capsular material. Seaweed alginates are (1-4) linked block copolymers of ß-D-mannuronic acid and a-L-guluronic acid (*7, 8*) while the bacterial polymers are generally randomly organized polymers (*9*). Most of the mannuronic acid residues in the bacterial alginates are mono-O-acetylated at the C-2 or C-3 positions, with some mannuronates being 2,3 di-O-acetylated (*10, 11*).

Alginates containing high proportions of mannuronate residues adopt flat ribbon-like 2-fold chain conformations in the solid state, similar to those found in ß-1,4 diequatorially linked polymers such as cellulose (*12, 13*), whereas polyguluronate rich polymers adopt a buckled 2-fold chain conformation (*14*). The mannuronate/guluronate (M/G) ratio's in alginates vary depending on the source of the polymer (*15, 16*). The M/G ratio, along with the sequence (block or random) influences the polymer's solution properties and the properties of any resultant gels.

Alginates with low M/G ratios generally produce low viscosity solutions and more brittle gels. Alginates with high M/G ratios produce solutions with higher viscosity and more pliable gels. Alginic acid polymers form interchain associations in the presence of cations (particularly calcium) producing hydrated gels. This ability to gel in the presence of cations has made this a widely used industrial polysaccharide. The total world production alginates and derivatives is not known but is reputed to be 15,000- 20,000 tons/ year. The affinity of alginates for metal ions vary.

The cation adsorption affinities of alginates from different alga strains has been studied (17-21). Alginate prepared from *Laminaria digitata* (rich in mannuronate residues) has a different affinity for divalent metals than alginates prepared from *Laminaria hyperborea* stipes (rich in guluronate residues) (22). The primary mechanism for metal adsorption is ion-exchange (23, 24) although covalent bonding also plays a role (25). The carboxylic groups are involved in this binding (26). Acetyl groups have a marked effect on the macromolecular properties of the alginates. This is partially demonstrated by the increased gel flexibility and water holding capacities of bacterial alginates relative to seaweed alginates (27). Acetylation of alginate modifies its ionization properties by reducing the net negative charge of the polymer (28). It does not significantly alter the molecular weight or polydispersity of alginate (29).

Bacterial alginates show a unique range of characteristics not found with seaweed alginates because of their variable M/G (mannuronic/guluronic) ratio's and acetylation. The bacterium *P. syringae* produces an alginate that has essentially all of its mannuronic acid residues acetylated. In this strain of bacteria, the acetylation process is independent of bacterial alginate biosynthesis (30), making possible its use for 2-O acetylating seaweed alginate (31). This report illustrates the use of immobilized cells of this microorganism for biomodification of commercial alginate, as well as demonstrating some of the property changes that can be achieved in this polymer.

Experimental

Bacterial strain and culture conditions. *P. syringae* subsp. *phaseolica* ATCC 19304 was selected because it produced a highly acetylated alginate. It was later determined that the acetylation mechanism(s) is independent of alginate production. Cultures were maintained at 4°C on a modified Dworkin and Foster (DF) agar (32). This medium contained (in grams per liter of deionized water): KH_2PO_4, 4.0; Na_2HPO_4, 6.0; $MgSO_4 \cdot 7 H_2O$, 0.2; $(NH_4)_2SO_4$, 0.5; NaCl, 0.4; KNO_3, 9.1; gluconic acid, 20.0; and agar, 15.0. Gluconic acid was sterilized separately and then aseptically added to the basal medium. The pH of the medium prior to sterilization was between 6.7 and 7.0. Broth cultures were prepared by inoculating *P. syringae* ATCC 19304 from agar slants into 50 ml of DF liquid medium in 250 ml Erlenmeyer flasks. Cultures were incubated for 30-35 hours at 30°C and 180 rpm on a NBS Model G25-KC rotary shaker, after which the absorbance at 660 nanometers was standardized to between 1.9 and 2.0. Bacterial alginate was isolated either from the "slime" removed from agar plates, containing 20 ml of either DF or nutrient agar (Difco Labs, Detroit, MI), or from 5 day shake flask culture broth's. Alginate production and the degree of acetylation were determined colorimetrically.

Alginate and acetyl quantitation. Alginate concentrations were determined as uronic acid according to the method of Blumenkrantz and Asboe-Hansen (*33*). The degree of acetylation was measured according to the method described by McComb and McCready (*34*). Prior to assay, samples were desalted either by dialysis against deionized water for 48 hours, at room temperature, or by passage through a Sephadex G-25 column. Sodium alginate isolated from *Macrocystis pyrifera* (Sigma Chemical Co., St. Louis, MO.) was used as a standard. The standard for acetyl was glucose pentaacetate (Sigma Chemical Co., St. Louis, MO).

Deacetylation of bacterial alginate. Bacterial alginate was deacetylated for comparison studies with seaweed alginate by dissolving it in deionized water at a concentration of approximately 1 mg/ml. Three volumes of this solution were mixed with one volume of 1 N sodium hydroxide solution. After incubation for 20 minutes at room temperature with gentle agitation, one volume of 1 N hydrochloric acid was added to neutralize the solution (final pH was about 7.0) and stop the reaction. The deacetylated bacterial alginate was dialyzed extensively against deionized water to remove residual salts.

Acetylation by immobilized *P. syringae* cells. A bioreactor containing carbon immobilized *P. syringae* ATCC 19304 cells was used for the continuous acetylation of seaweed alginate (*30*). The feed was 1.5 g/l *Macrocystis pyrifera* alginate and 1.0%(w/v) gluconic acid in 0.01 M phosphate buffer (pH 6.0). Temperature was maintained at 25°C and aeration was 0.4 standard liters per minute (SLPM) with sterile air. The reactor vessel was a 700 ml Kontes Airlift Bioreactor (Kontes Life Science Products, Vineland, NJ) containing 25 g of carbon catalyst with a working volume of 500 ml.

Purification of acetylated alginates. Acetylated alginate was mixed with an equivalent volume of isopropanol. This mixture was incubated for 12 hours at room temperature and then centrifuged at 8,000 x g for 40 minutes in a Sorval Superspeed Model RC-5B centrifuge (DuPont Co., Wilmington, DE). The precipitate was washed with isopropanol and redissolved in deionized water. It was dialyzed against 500 volumes of deionized water for 48 hours to remove low molecular weight compounds and then concentrated in a Büchi Model R110 rotary evaporator (Büchi Lab., Flawil, Switzerland). After concentration, it was freeze-dried and stored.

Nuclear Magnetic Resonance (NMR) spectra of alginates. NMR spectroscopy was used to confirm of acetylation seaweed alginate and position of acetylation. All samples for NMR were deuterated three times by evaporation under reduced pressure with 0.5 ml of D_2O (Sigma Chemical Co., St. Louis, MO). ^1H-NMR spectra of each sample were obtained using a Bruker WM-400 NMR spectrometer (Bruker

Instrument Co., Germany) operating in the Fourier transform mode. The sample volume for ^{1}H-NMR spectra was 0.5 ml. The alginate concentrations were 60 mg/ml, except for the bacterial alginate which was 40 mg/ml. The operating temperature was 70°C. Each ^{1}H-NMR spectra was scanned for 1 hour.

Viscosity. The effects of acetylation on the solution viscosity of alginates were determined by the method of Allison and Matthews (*35*) using a simple U-shaped capillary viscometer designed for small volume samples. The time taken for the adjusted concentration of each alginate sample to fall a fixed distance under gravity at a constant temperature was expressed as comparative viscosity (N/N_0), where N is the time elapsed for the alginate solutions to fall a fixed distance and N_0 the time for deionized water to fall that same distance. The concentration of all samples tested was 400 mg/ml. The volume of each sample was 2.5 ml. Measurements were repeated 5 times.

Molecular weights. Polysaccharide molecular weights were determined by gel permeation chromatography (GPC) of alginate [100 mg/ml (w/v)] in deionized water. Alginate sizes and polydispersity indices were determined by measurement of multiangle light scattering intensities using a DAWN-Photometer (Wyatt Technology, Santa Barbara, CA). The DAWN GPC detector measures the scattering intensities of a sample at 15 different angles and transmits the data to a computer for digital conversion and subsequent processing under control of the ASTRA‰ (or ASTRA 202) software (Wyatt Technology, Santa Barbara, CA). Acetone and cyclohexane were used for instrument calibration. Concentrations were obtained from a calibrated Waters Model 410 Differential Refractometer (Millipore Corp., Milford, MA). Alginate sizes were determined using: T10, T40, and T500 dextrans (Pharmacia Co., Uppsala, Sweden) as external standards. Sample injection volumes were 100 ml and the GPC column was an Ultrahydrogel Linear column (Waters, Millipore Corp., Milford, MA). The running buffer was 0.1 M NaNO$_3$ and the temperature was 45°C.

Precipitation of alginates by metal ions. The relative precipitation of alginates by cations were compared by determining the concentration of cation required to precipitate 50% of a standard concentration alginate solution, 400 mg/ml (w/v). Metal salts were dissolved in deionized water at concentrations ranging up to 100 mM. The metal ions tested were: Ca^{2+}, Cs^{1+}, Co^{2+}, Fe^{3+}, Pb^{2+}, Mg^{2+}, Rb^{1+}, Sr^{2+}, and U^{6+}. All metal salts were obtained from Sigma Chemical Co., St. Louis, MO, except for uranyl acetate (Eastman Kodak Co., Rochester, NY). Four volumes of each alginate solution were mixed with one volume of each metal solution. The solutions were incubated for 12 hours at room temperature to allow for maximum complex formation. They were then centrifuged (18,000 x g for 60 minutes) in a Sorval Superspeed Model RC-5B centrifuge. The supernatants were separated and the concentration of residual alginate in each supernatant was measured chemically. The concentrations of alginate precipitated were calculated by difference.

Results

Separation of acetylation. Alginate production patterns by *P. syringae* ATCC 19304 resemble those for mixed metabolites (neither primary or secondary), whereas acetylation of the polymer shows a primary metabolite pattern (Figure 1). When *P. syringae* was cultured in the presence of a high concentration of seaweed alginate (800 ug/ml) with increasing concentrations of a carbon-energy source, gluconic acid, alginate acetylation increased. There was no increase in the total uronic acid present in the solution (Figure 2) indicating inhibition of bacterial alginate biosynthesis, but not of acetylation. This implied that the acetylation mechanism for alginate in this bacteria is not tightly linked to biosynthesis of the polysaccharide. Of the two dozen *P. syringae* strains that were tested, this phenomenon was only observed with this strain. Separation of biosynthesis and acetylation of alginate allowed this strain of microorganism to be used as a biocatalyst for selective acetylation of seaweed alginate.

Acetylation of seaweed alginate by *P. syringae* cells. Resting cell suspensions were able to acetylate seaweed alginate regardless of growth phase of the cells when harvested. The highest level of polysaccharide modification was achieved if mid-log phase cells were used. Aeration was required for maximum polymer acetylation and an added carbon source (gluconic acid) was required to maintain acetylation. In the presence of seaweed alginate and gluconic acid, the microorganism did not produce a bacterial alginate.

A fluidized bed reactor was constructed using a granular activated carbon charged with *P. syringae* cells. Granular activated carbon provided a matrix for maintenance of high numbers of the microorganism. The reactor was fed a buffered alginate solution containing gluconic acid. After five days of operation the number of cells on the surface of the carbon particles increased 25 fold, demonstrating that the culture was not static, but rather slowly growing (*30*). The reactor apparently operated like a low dilution rate, stabilized, continuous culture apparatus, allowing a slow rate of growth. Acetylation of alginate reached a maximum (90% of the available mannuronate residues were acetylated) three days after startup and then the rate of reaction decreased linearly (Figure 3). The $T_{1/2}$ of this system was 6.5 days at a flow rate of 0.02 h^{-1}. A constant level of acetylation could be maintained by varying the feed rate.

The presence of acetyl groups in the seaweed polymer was confirmed by NMR studies. Acetylation of the seaweed alginate produced a new, distinct, signal between 2.0 and 2.5 PPM in the ^{1}H-NMR spectrum. This signal was absent in the ^{1}H-NMR spectrum of the native seaweed polymer (*36*). The signal location between 2.0 and 2.5 PPM is characteristic of methyl groups in an acetylated region. The partly overlapping peaks of acetyl protons in this region suggests the presence of either di-acetylated units and/or two mono-acetylated units (*37*). Bacterial alginate shows the same signal between 2.0 and 2.5 PPM on ^{1}H-NMR spectrum. This signal was absent in the ^{1}H-NMR spectrum of deacetylated bacterial alginate.

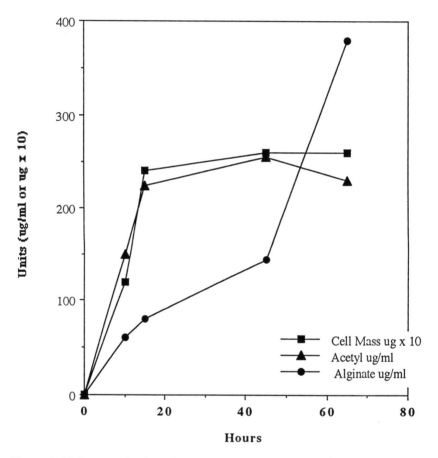

Figure 1. Alginate production ($\blacklozenge$), acetyl on polysaccharides ($\blacktriangle$) and growth ($\blacksquare$) of *P. syringae* ATCC 19304. Cell mass is given as 10 x ug/ml, all other units are ug/ml. Adapted from ref. 30.

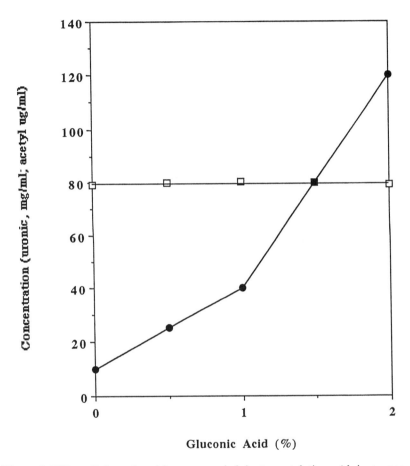

Figure 2. Effect of gluconic acid on seaweed alginate acetylation. Alginate was supplied to immobilized cells at a concentration of 800 ug/ml. Samples were taken 5 days post inoculation. Alginate acetylation (●) and total uronic acid concentrations (□) were determined as a function of gluconic acid concentration in the feed. Adapted from ref 30.

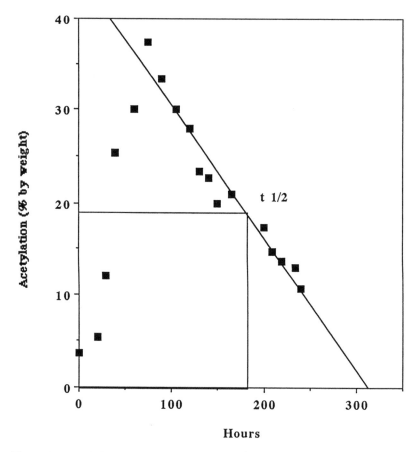

Figure 3. Acetylation of Macrocystis alginate by immobilized *P. syringae* cells. The feed contained 1.5 g/l of alginate and 1.0% gluconic acid in 0.01 M phosphate buffer (pH 6.0). The feed rate was 0.02 h⁻¹, temperature 25°C and aeration 0.4 SLPM.

Viscosity Effects. Solution viscosity is a function of molecular size and rigidity of the solute, as well as environmental factors such as temperature. A linear increase in alginate concentration produced a non-linear increase in the comparative viscosity (N/N_0). Seaweed alginate showed small deviations from Newtonian flow with increasing concentration. Acetylation did not change this, but did increase the observed viscosity by about 15%. Acetylation directly affected the flow dynamics, a 30% acetylated sample showed an increase in viscosity of 8% over the control polymer between 20 and 52 °C (Figure 4). These changes are probably due to the increase in size of the polymer because of the acetylation (Table I) , rather than to increased chain-chain interactions because of the acetylation.

Table I Molecular Sizes of Alginates

Alginate sample	$M_n{}^a$ (x 10^4)	$M_w{}^b$ (x 10^4)	$M_w/M_n{}^c$
Seaweed	1.4	4.7	3.36
Acetylated seaweed	1.6	5.2	3.25
Bacterial	4.3	12.7	2.95
Deacetylated bacterial	3.8	11.9	3.13

[a]Number average molecular weight. [b]Weight average molecular weight. [c]Polydispersity.

Source: Adapted from Ref. 36

Gelation with Cations. Precipitation of the various alginates with metal ions was used as a measure of affinity of the polymer for the cation. The relative ability of various cations to precipitate alginates are reported as $P_{1/2}$ values where the $P_{1/2}$ is defined as the concentration of metal ion (mM) required to precipitate 50% of the polymer from a 400 ug/ml (w/v) solution. In most cases the acetylated polymer required significantly higher cation concentrations for precipitation than the equivalent concentration of non-acetylated polymer. There was a significant reduction in the ability of cobalt, calcium and strontium ions to precipitate acetylated alginate, as well as a major loss ability of the polymer to interact with either gold or ferrous ions. This is most likely due to steric hindrance of cation binding sites by the acetyl groups. The exception to this pattern was ferric ion where the acetylated polysaccharide showed a 3X higher affinity for this metal than did the non-acetylated polymer (Table II).

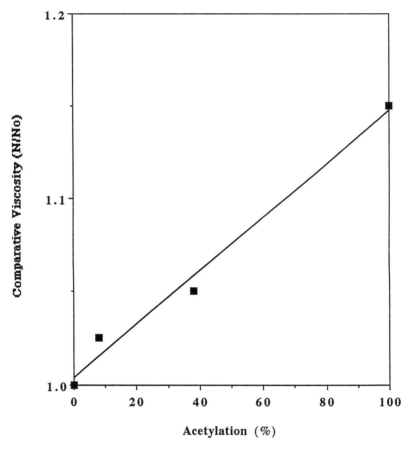

Figure 4. Effect of acetylation on the comparative viscosity of Macrocystis alginate. Reproduced with permission from ref. 27.

Table II. The effect of acetylation on the precipitation of alginates by cations expressed as a $P_{1/2}$[a]

Ion	Alginate	Acetylated
U^{6+}	0.6	0.8
Pb^{2+}	1.5	5.0
Ca^{2+}	2.0	15.0
Fe^{3+}	6.5	2.0
Co^{2+}	9.5	50.0
Sr^{2+}	10.0	19.0
Fe^{2+}	7.5	25<
Au^{3+}	10.0	50<

a. The concentration of cation in mM that precipitates 50% (w/v) of the alginate from a 400 ug/ml alginate solution.
Source: Adapted from Ref. 36

Discussion

Acetylation alters both the solution viscosity and some of the cation gelling characteristics of seaweed alginate. It produced a 10.6% increase in the M_W of the polymer. This translated into a 4% increase in comparative viscosity for the acetylated polymer at $30^{\circ}C$. Bacterial alginate solutions show a much greater viscosity difference between the acetylated and deacetylated polymers at equivalent concentrations. At $30^{\circ}C$ the comparative viscosity of the acetylated bacterial polymer is 38% higher than the deacetylated polymer (*36*). The difference between the two polysaccharides may be in part due to the increase in average molecular weight, but is more likely due to the higher degree of acetylation and the more extended ribbon-like tertiary structure of the bacterial polymer.

The acetylated alginates showed major shifts in their ability to gel with cations, generally acetylation decreased gelling ability. It is apparent that acetylation decreases a polymers ability to gel with those divalent cations which normally have high affinity for alginate. The high affinity of seaweed alginate for calcium ion is due to the structural characteristics of the polymer; polyguluronate residues, hydroxyl groups and carboxyl groups of polyguluronate residues, and net negative charge, as well as the molecular size and net charge of the calcium ion, all of which are implicated in this interaction (*13, 14, 38*). The acetylation of the seaweed alginate probably disturbs all the basic relationships that exist between the polymer and calcium by modifying the ionization properties of the polymer and sterically hindering the binding of the ions to the polymer (*39, 40*). The only increase observed in gelling capability was with ferric ions. This may have major physiological significance for bacterial systems where the organisms normally live in an iron deficient environment.

Because of a higher viscosity and a lower affinity for calcium ions, acetylated alginates can be favorably substituted for nonacetylated alginates when used as emulsifiers, stabilizers, and gelling agents in many industrial applications. The lower affinity of acetylated alginate for calcium ions confers a more soluble state on the polymer in aqueous solution. Acetylated alginate thus becomes a more desirable emulsifier and stabilizer. Higher viscosity solutions can be made with lower concentrations of acetylated alginate reducing the amount of polymer required in any given application.

References

1. Larsen, B., Haug, A., Carbohydr. Res., **1971**, *32*, 217-225.
2. Hacking, A. J., Taylor, I. W. F., Jarman, T. R., Govan, J. R. W., J. Gen. Microbiol., **1983**, *129*, 3473-3480.
3. Linker, A., Jones, R. S., J. Biol. Chem., **1966**, *241*, 3845-3851.
4. Banerjee, P. V., Vanags, R. I., Chakrabarty, A. M., Maitra, P. K.; J. Bacteriol., **1983**, *155*, 238-245.
5. Fyfe, J. A. M., Govan, J. R. W., J. Gen. Microbiol., **1980**. *119*, 443-450.
6. Fett, W. F., Osman, S. F., Fishman, M. L., Siebles III, T. S., Appl. Environ. Microbiol., **1986**. *52*, 466-473.
7. Lin, T., Hassid, W. Z., J. Biol. Chem., **1966**, *241*, 5284-5298.
8. Penman, A., Sanderson, G. R., Carbohydr. Res., **1972**, *25*, 273-282.
9. Davidson, J. W., Sutherland, I. W., Lawson, C. I., J. Gen. Microbiol., **1977**. *98*, 603-606.
10. Sherbrock-Cox, V., Russell, N., Gacesa, P., Carbohydr. Res., **1984**, *135*, 147-154.
11. Skjak-Braek, G., Larsen, B., Grasdalen, H., Carbohydr. Res., **1985**, *145*, 169-174.
12. Atkins, E. D., Nieduszynski, I. A.; Biopolymer, **1973**. *12*, 1865-1878.
13. Rees, D. A., Biochem. J., **1977**, *126*, 257-273.
14. Atkins, E. D., Nieduszynski, I. A.; Biopolymer, **1973**, *12*, 1879-1887.
15. Haug, A., Larsen, B., Acta. Chem. Scand., **1962**, *16*, 1908-1918.
16. Haug, A., Smidsrod, O., Acta. Chem. Scand., **1965**, 19, 1221-1226.
17. Darnall, D. W., Greene, B., Henzl, M. T., Hosea, J. M., McPherson, R. A., Sneddon, J., Alexander, M. D. Environ. Sci. Technol., **1986**, *20*, 206-208.
18. Holan, Z. R., Volesky, B., Biotechnol. Bioeng., **1994**, *43*, 1001-1009.
19. Nakajima, A., Horikoshi, T., Sakaguchi, T., Eur. J. Appl. Microbiol. Biotechnol., **1982**, *16*, 88-91.
20. Torresday, J. L., Darnall, D. W., Wang, J., Anal. Chem., **1988**, *60*, 72-76.
21. Volesky, B., Prasetyo, I., Biotech. Bioeng., **1994**, *43*, 1010-1015.
22. Haug, A., Acta. Chem. Scand., **1961**, *15*, 1794-1795.
23. Kohn, R., Pure. Appl. Chem., **1975**, *42*, 371-397.
26. Torresday, J. L., Hapak, M. K., Hosea, J. M., Darnall, D. W., Environ. Sci. Technol., **1990**, *24*, 1372-1378.

24. Kuyucak, N., Volesky, B., Biotechnol. Bioeng., **1989**, *33*, 823-831.
25. Watkins, W., Elder, R. C., Greene, B., Darnall, D. W., Inorg. Chem. 1987, *26*, 1147-1151.
27. Day, D. F. , Ashby, R. D., Proceedings of the American Chemical Society Division of Polymeric Materials: Science and Engineering. **1995**, *72*, 1 41- 142
28. Dentini, M., Crescenzi, V., Blasi, D., Int. J. Biol. Macromol., **1984**. *6*:93-98.
29. Skjak-Braek, G., Zanetti, G. F., Paoletti, S., Carbohydr. Res., **1989**, *185*, 131-138.
30. Lee, J., Day, D., Appl. Environ. Microbiol., **1995**, *61*, 650-655.
31. Day D.F., Lee, J., US Patent 5,308,761, **1994**.
32. Dworkin, M., Foster, J. W., J. Bacteriol., **1958**. *75*, 592-603.
33. Blumenkrantz, N., Asboe-Hansen, G., Anal. Biochem., **1973**. *54*, 484-489.
34. McComb, E. A., McCready, R. M., Anal. Chem., **1957**, *29*, 819-821.
35. Allison, D. G.; Matthews, M. J. *J. Appl. Bacteriol.*, **1992**. *73*, 484 -
36 Lee, J. W., Ashby, R. D., and Day, D. F.Carbohydrate Polymers, **1996**. 29, 337-345
37. Skjak-Braek, G. Paoletti, S. and Gianferrara., Carbohydr. Res **1989**, *185*, 119-129.
38. Nilsson, S., Biopolymer, **1992**, *32*, 1311-1315.
39. Morris, E. R., Rees, D. A., Thom, D., Carbohydr. Res., **1978**, *66*, 145-154.
40. Morris, E. A., Rees, D. A., J. Mol. Biol., **1980**, *138*, 363-374.

Chapter 11

Enzymatic Modification of Chitosan by Tyrosinase

Joseph L. Lenhart[1], Mahesh V. Chaubal[1], Gregory F. Payne[1],
and Timothy A. Barbari[2]

[1]Department of Chemical and Biochemical Engineering and Center
for Agricultural Biotechnology, University of Maryland Baltimore County,
5401 Wilkens Avenue, Baltimore, MD 21228
[2]Department of Chemical Engineering, The Johns Hopking University,
3400 North Charles Street, Baltimore, MD 21218–2694

Mushroom tyrosinase was used to enzymatically modify thin films of
chitosan polymer by attaching different phenolic functionalities. The
enzyme catalyzes the oxidation of the phenol to an o-quinone, which is
then able to freely diffuse to and covalently bond with the nucleophilic
amine groups of chitosan. The tyrosinase was shown to react with a
wide variety of phenols, including: phenol, catechol, caffeic acid, and
L-dihydroxyphenylalanine, while UV-Visible spectrophotometry of thin
films was used to demonstrate reaction between the quinone product
and chitosan. Since chitosan is a natural, biodegradable polymer, and
tyrosinase reacts with a large range of phenols, the potential exists to
develop useful, environmentally friendly polymers by the enzymatic
attachment of these different phenolic groups to the chitosan.

There are several approaches to using biological catalysts in polymer manufacturing.
One approach is to employ living systems for synthesizing the monomer. Two
examples of such monomer synthesis are the fermentation production of lactic acid (for
the synthesis of polylactides), and the biotransformation of acrylonitrile to acrylamide
(*1*). A second approach is to use living systems directly to synthesize a biopolymer.
For example fermentation is routinely used to produce proteins and polysaccharides
and more recently for polyhydroxyalkanoates (*2,3*). A final approach for using
biological catalysts in polymer synthesis is the use of enzymes. Polycondensation
reactions can be catalyzed if hydrolytic enzymes are incubated in non-aqueous
environments such that the hydrolytic reactions are reversed. Studies on enzyme-
catalyzed polycondensation reactions have involved lipase (*4-6*), hydrolase (*7*), and
protease (*8*) enzymes. Also, polyphenolic polymers can be synthesized by free-radical
reactions catalyzed by peroxidase enzymes (*9-12*).

Tyrosinase-Catalyzed Polymer Modification

In our work we are examining an enzymatic method for modifying pre-formed chitosan
polymers. Chitosan can be obtained by extracting chitin from the shells of crustaceans
(e.g. crabs, shrimp and lobster) and then de-acetylating the chitin to form chitosan.

188

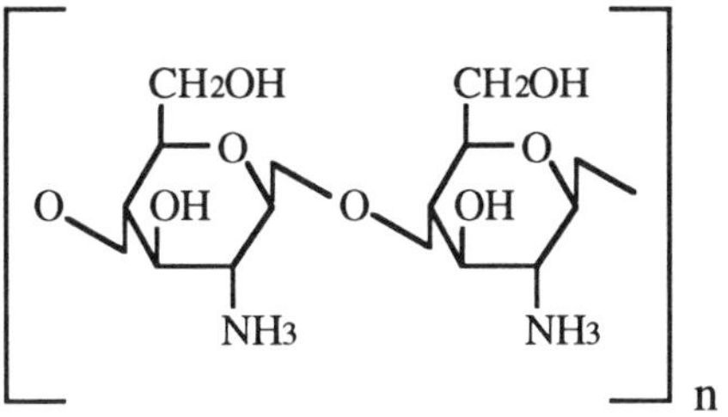

Chitosan

The Chitosan Polymer. The advantages of using chitosan are: (i) the polymer is pre-formed and difficulties in polymer synthesis can be avoided; (ii) the polymer can be obtained from waste products (i.e. shells) of the seafood industry; (iii) the polymer is renewable and biodegradable, and hopefully after modification would remain biodegradable. Unfortunately, because of the amounts of acids and bases required for de-mineralization, de-proteination and de-acetylation of crustacean shells, current production processes for chitosan yield relatively expensive polymer ($7-$27/lb) (*13*).

One important feature of chitosan is that the polymeric backbone contains primary amine groups. This amine functionality can be exploited at low pH to confer polyelectrolyte properties while at higher pH's the neutral amine is nucleophilic and can undergo a variety of reactions. Although amine functionality is desirable for various applications, there are few amine-rich natural polymers, while synthetic amine-containing polymers often require hazardous reactions (e.g. chloromethylation).

Tyrosinase Catalyzed Reaction Followed by Chitosan Adsorption. The enzyme we are using is a polyphenol oxidase, tyrosinase, which catalyzes the oxidation of phenols to ortho quinones (reaction 1). These quinones are unstable and react with the nucleophilic group of chitosan (reaction 2) to form the modified chitosan polymer.

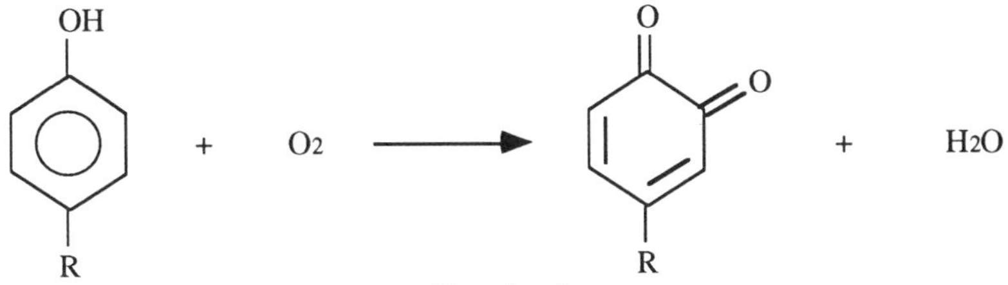

Reaction 1

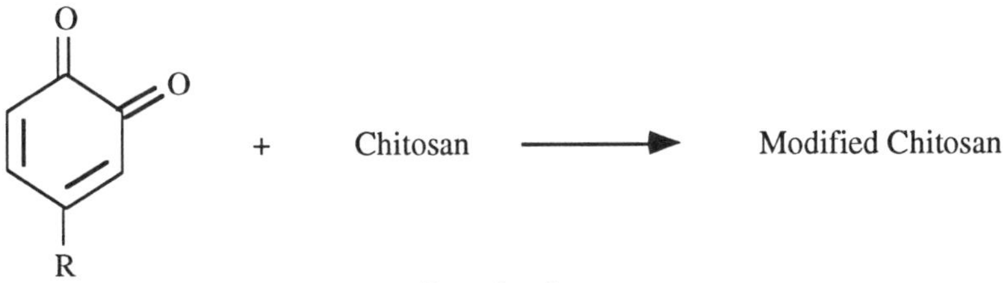

Reaction 2

Tyrosinase is an unusual enzyme because it actually catalyzes two separate reactions - the hydroxylation of monophenols to diphenols and the further oxidation of the diphenol to o-quinones (reaction 3).

Reaction 3

Since both tyrosinase-catalyzed reactions consume molecular oxygen, the progress of the reaction can be monitored simply by following changes in the dissolved oxygen concentration (*14*). For example, Figure 1 shows that both phenol (5 mM) and the ortho diphenol, catechol (5 mM), can be oxidized by mushroom tyrosinase (70 Units/mL). However, the oxidation of catechol is considerably more rapid (*15*).

In addition to "activating" phenols by conversion to reactive quinone products, there are several other attractive features of the tyrosinase-catalyzed reaction for polymer modification. First, tyrosinase utilizes molecular oxygen as the oxidant and does not require complex cofactors such as NADH. Second, tyrosinase is relatively selective (i.e. chemoselective) for reactions with phenols. For instance tyrosinase was observed to oxidize the methyl-substituted phenol, cresol, while the aromatic ether, anisole, and the aromatic alcohol, benzyl alcohol, were not substrates for tyrosinase (*16*). Third, although limited to catalyzing reactions with phenols, tyrosinase has a broad substrate range for phenols which provides an opportunity for modifying the chitosan polymer with a variety of phenolic functionalitites.

A final feature of the enzymatic modification approach, is that the quinone product of the first reaction dissociates from the enzyme and freely diffuses to the amine site on the chitosan polymer. This eliminates steric limitations which are commonly encountered when enzyme-bound intermediates must be transferred to a polymer, as is often the case with enzyme-based condensation strategies.

Chitosan Film Preparation, Modification, and Analysis. In initial studies, we used UV/Vis spectrophotometry to demonstrate that tyrosinase could be used to modify chitosan surfaces (*17*). A schematic diagram of the chitosan film preparation and UV-visible scanning procedure is given in Figure 2. In this experiment, 2 g of chitosan was dissolved in 100 mL of 8 v/v% acetic acid, the viscous acidic solution of chitosan was cast onto glass slides, and the slides were immersed in a caustic solution to gel the chitosan into films. After washing the films with distilled water and a 50 mM phosphate buffer, they were then incubated with solutions containing various components (either 5 mM of catechol, 70 Units/mL mushroom tyrosinase, or both), washed with the phosphate buffer, and the UV-visible spectra was measured. The chitosan film immersed in only the phosphate buffer was scanned as the reference. As shown in Figure 3, when a film was incubated with either catechol or tyrosinase alone, the gel remained colorless and the UV-visible spectra was similar to that of the reference. However, when the film was incubated with both tyrosinase and catechol for 30 minutes, the film was observed to become brown and became UV-absorbing (Figure 3). Similar results have been observed with other phenolic substrates (*17*), and indicate that the reaction products from the tyrosinase-catalyzed reaction are readily bound to chitosan. Although we have not yet characterized the chemistry involved in

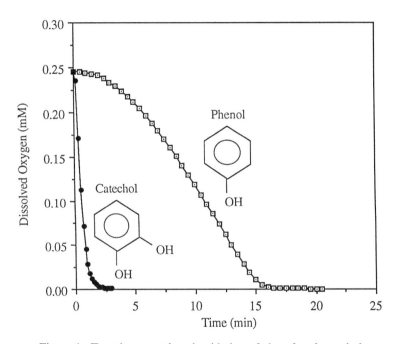

Figure 1. Tyrosinase catalyzed oxidation of phenol and catechol.

I. Chitosan Film Preparation

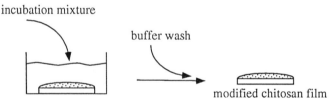

viscous chitosan solution → glass slide → slide coated with chitosan solution

caustic treatment → slide coated with thin chitosan gel film

water and phosphate buffer wash → slide coated with neutralized chitosan film

II. Chitosan Film Modification

incubation mixture

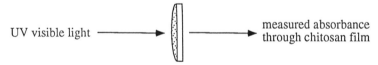

buffer wash → modified chitosan film

III. UV/Visible Analysis of the Film

UV visible light ⟶ ⟶ measured absorbance through chitosan film

Figure 2. Schematic of chitosan film preparation, modification, and UV-Visible scanning of the modified film.

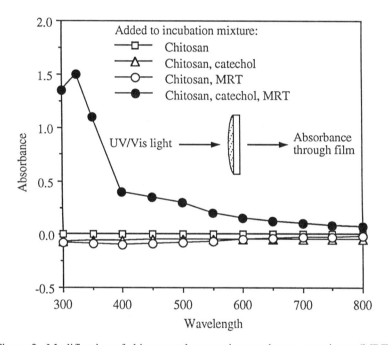

Figure 3. Modification of chitosan polymer using mushroom tyrosinase (MRT) and catechol.

the quinone-chitosan binding, related reactions involving the binding of amines to the phenolic fractions of soil (18,19), and reactions reported for quinone-amine polymerization (20) suggest that binding may involve either a Michael's-type adduct or a Schiff base. We are currently investigating the chemistry involved in quinone-chitosan binding.

Potential Modifications of Chitosan

In addition to demonstrating the concept, our initial studies have suggested the potential for using tyrosinase for altering the physical/chemical properties of a chitosan surface. For instance, the ability to enzymatically modify chitosan by the addition of tert-butylcatechol (17) suggests the potential for using phenols with large non-polar functional groups (e.g. nonylphenol) to produce hydrophobically modified chitosan. The success of such an approach depends on the ability to overcome the low water solubility of such a hydrophobic reactant. Important to note however, is that the tyrosinase enzyme is known to be capable of functioning in non-aqueous environments (21-23).

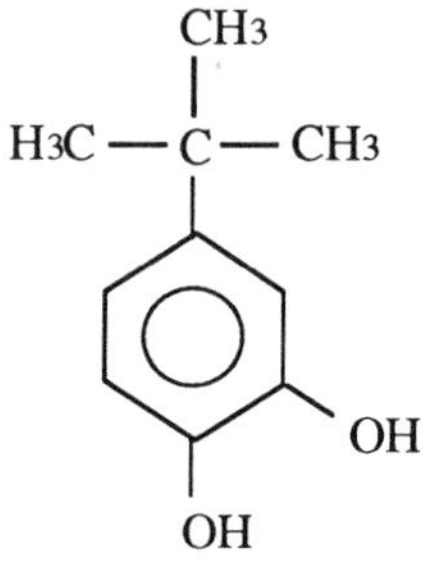

$$CH_3$$
$$H_3C - C - CH_3$$

OH

OH

tert-butylcatechol

In addition to altering the hydrophobicity of chitosan, previous results have shown that the addition of the acidic phenol, *p*-hydroxyphenoxyacetic acid yields a modified chitosan with acidic properties (17).

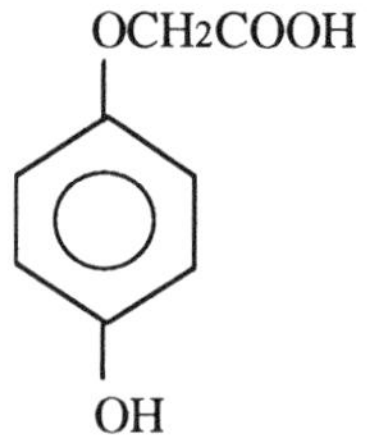

$$OCH_2COOH$$

OH

p-hydroxyphenoxyacetic acid

The potential exists for adding functional groups to chitosan which can undergo free-radical reactions. For instance, Figures 4 and 5 show that 1.25 mM caffeic acid can be enzymatically oxidized by tyrosinase (70 Units/mL) and added to a chitosan film. We are currently exploring the whether caffeic-acid-modified chitosan can undergo free-radical catalyzed reactions through the double bond of the acrylate group. If this group can enter into free-radical catalyzed reactions, then it may be possible to

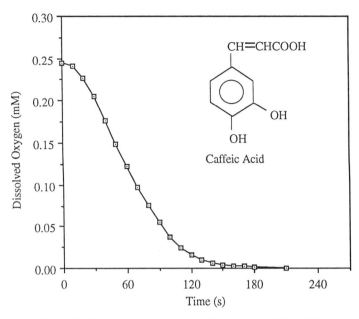

Figure 4. Tyrosinase catalyzed oxidation of caffeic acid.

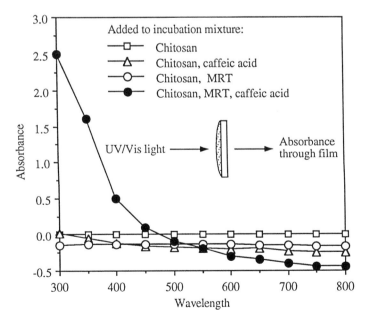

Figure 5. Modification of chitosan polymer using mushroom tyrosinase (MRT) and caffeic acid.

use such reactions to cross-link the chitosan chains, or graft monomers and polymers onto chitosan.

A final opportunity is suggested by Figures 6 and 7 which shows that 5 mM dihydroxyphenylalanine (DOPA) can be oxidized by tyrosinase (70 Units/mL) and the product can be bound to chitosan. Since it is known that tyrosinase can oxidize tyrosine and even tyrosine residues of peptides (24-26), the ability to bind the reaction product to chitosan suggests the potential for adding amino acids, peptides and proteins to the chitosan polymer. Potential advantages of a tyrosinase-catalyzed peptide grafting method is the modest conditions required for grafting and the specificity of the enzyme for the tyrosine residue of the peptide.

Conclusions

The oxidation of phenols by polyphenoloxidase enzymes (e.g. tyrosinase) is known to generate reactive compounds (e.g. *o*-quinones) which can undergo a variety of reactions with nucleophiles. These reactions have been well studied for the synthesis of melanins (27,28), the browning of food (29), and the irreversible binding of amine-containing pesticides to soil (30-32). Reactions between enzyme-generated quinones and nucleophiles have also been suggested as a potential defense mechanism employed by plants to discourage insect feeding (33-36). The above-mentioned studies have demonstrated that the chemistries involved in these reactions can be varied and complex.

We are exploring whether quinone-generating enzymatic reactions can be exploited for the functionalization of nucleophilic polymers (especially for the amine-rich natural polymer chitosan). The attraction of exploiting reactions 1 and 2 for

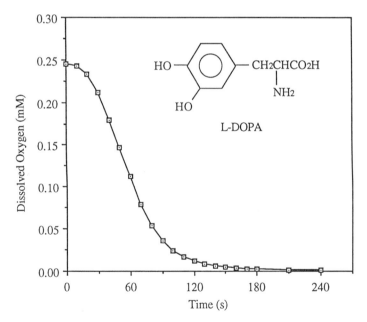

Figure 6. Tyrosinase catalyzed oxidation of L-Dihydroxyphenylalanine (L-DOPA)

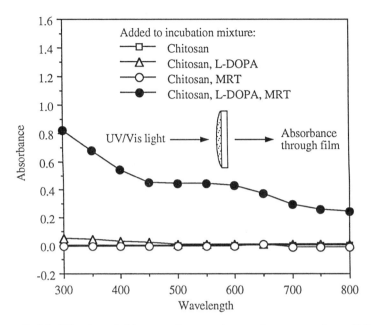

Figure 7. Modification of chitosan polymer using mushroom tyrosinase (MRT) and L-Dihydroxyphenylalanine (L-DOPA).

polymer modification is that the reactions are simple, rapid and occur under mild conditions (aqueous solutions at room temperature and at neutral pH). Thus the enzyme offers the potential for functionalizing the polymers under environmentally more-friendly conditions compared to many alternative synthetic methods. Also, the ability to enzymatically "activate" the phenol into the reactive quinone suggests the potential for the *in situ* generation of the group to be added to the polymer. *In situ* generation of the reactive species may offer significant safety advantages compared to typical synthetic methods which require the synthesis, transport and storage of reactive functionalizing groups (e.g. ethylene oxide, epichlorhydrin or phosgene).

Although our initial studies (*17*) and the data presented here suggest the potential of tyrosinase-catalyzed reactions for polymer modification, much work remains before reactions 1 and 2 can be exploited practically. First, it is necessary to characterize the chemistry involved in binding of the quinones to chitosan. Second, to understand how the enzymatically-modified chitosan can be exploited practically, it will be necessary to determine how the polymer properties are altered by modification.

Acknowledgments

This work was partially supported by grants from the National Science Foundation (BCS-9315449) and Maryland Sea Grant (NA-46R60091). Laboratory assistance from Amy Yorks is also appreciated.

Literature Cited

1. Kobayashi, M., Nagasawa, T., Yamada, H. *Trends Biotechnol.* **1992**, *10*, 402.
2. Lee, S.Y. *Biotechnol. Bioeng.*, **1996**, *49*, 1.
3. Doi, Y. *Microbial Polyesters*. VCH Publishers Inc.: New York, NY, 1990.

4.   Geresh, S., Gilbao, Y. *Biotechnol. Bioeng.* **1990**, *36* , 270.
5.   O'Hagan, D., Zaidi, N.A. *Polymer.* **1994**, *35*, 3576.
6.   Brazwell, E.M., Filos, D.Y., Morrow, C.J. *J. Polym. Sci. Part A. Polym. Chem.* **1995**, *33*, 89.
7.   Abramowicz, D.A., Keese, C.R. *Biotechnol. Bioeng.* **1989**, *33*, 149.
8.   Patil, D.R., Rethwisch, D.G., Dordick, J.S. *Biotechnol. Bioeng.* **1991**, *37*, 639.
9.   Dordick, J.S., Marletta, M.A., Klibanov, A.M. *Biotechnol. Bioeng.* **1987**, *30*, 31.
10.   Rao, A.M., Vijay, J.T., Gonzalez, R.D., Akkara, J.A., Kaplan, D.L. *Biotechnol. Bioeng.* **1995**, *28*, 531.
11.   Ayyagari, M.S., Marx, K.A., Tripathy, S.K., Akkara, J.A., Kaplan, D.L. *Macrmolecules* . **1995**, *28* , 5192.
12.   Bruno, F.F., Akkara, J.A., Kaplan, D.L., Sekher, P., Marx, K.A., Tripathy, S.K. *Ind. Eng. Chem. Res.* **1995**, *34* , 4009.
13.   Mathur, N.K., Narang, C.K. *J. Chem. Ed.* **1990**, *67*, 938.
14.   Mayer, A.M., Harel, E., and Ben-Shaul, R. *Phytochem.* **1966**, *5*, 783.
15.   Sun, W.-Q., Payne, G.F., Moas, M.S.G.L., Chu, J.H., Wallace, K.K. *Biotechnol. Progr.* **1992**, *8*, 179.
16.   Payne, G.F., Sun, W.-Q., Sohrabi, A. *Biotechnol. Bioeng.* **1992**, *40*, 1011.
17.   Payne G. F., Chaubal M. V., Barbari T. A. *Polymer.* **1996**, In Press.
18.   Parris, G.E. *Environ. Sci.Technol.*. **1980**, *14*, 1099.
19.   Tatsumi, K., S.-Y. Liu, and J.M. Bollag.*Wat.Sci .Technol.* **1992**, *25*, 57.
20.   Nithianandam, V.S., Erhan, S. *Polymer.* **1991**, *32*, 1146.
21.   Kazandjian, R.Z., and Klibanov, A.M. *J. Am. Chem. Soc.* **1985**, *107*, 5448.
22.   Estrada, P., Baroto, W., Castillon, M.P., Acebal, C., Arche, R. *J. Chem. Tech. Biotechnol.* **1993**, *56*, 59.
23.   Burton, S.G., Duncan, J.R., Kaye, P.T., Rose, P.D. *Biotechnol. Bioeng.* **1993**, *42*, 938.
24.   Ito, S., Kato, T., Shinpo, K., and Fujita, K. *Biochem. J.* **1984**, *222*, 407.
25.   Marumo, K., Waite, J.H. *Biochim. Biophys. Acta* . **1986**, *872*, 98.
26.   Rosei, M.A., Mosca, L. Coccia, R., Blarzino, C., Musci, G. De Marco, C. *Biochim. Biophys. Acta* . **1994**, *1199*, 123.
27.   Mason, H.S., Wright, C.I. *J. Biol. Chem.* **1949**, *180*, 235.
28.   Aroca, P., Solano, F., Salinas, C., Garcia-Borron, J.C., Lozano, J.A. *Eur. J. Biochem.* **1992**, *208*, 155.
29.   Freidman, M. *J. Agric. Food Chem.* **1996**, *44*, 631.
30.   Liu, S.-Y., Minard, R.D., Bollag, J.-M. *J. Agric. Food Chem.* **1981**, *29*, 253.
31.   Nannipieri, P. Bollag, J.-M. *J. Environ. Qual.* **1991**, *20*, 510.
32.   Bollag, J.-M. *Environ. Sci. Technol.* **1992**, *26* , 1876.
33.   Vaughn, K.C., Lax, A.R., Duke, S.O. *Physiol. Plant.* **1988**. *72*, 659.
34.   Felton, G.W., Donato, K., Del Vecchio, R.J., and Duffey, S.S. *J. Chem. Ecol.* **1989**, *15* , 2667
35.   Felton, G.W., Donato, K.K., Broadway, R.M., and Duffey, S.S. *J. Insect Physiol.* **1992**, *38* , 277.
36.   Constabel, C.P., Bergey, D.R., Ryan, C.A. *Proc. Natl. Acad. Sci.* **1995**, *92*, 407.

Chapter 12

Chemoenzymatic Synthesis and Modification of Monomers and Polymers

Helmut Ritter

Macromolecular Chemistry, University of Wuppertal, FB 9, Gausstrasse 20, D–42097 Wuppertal, Germany

Abstract
The paper describes the use of enzymes for the synthesis of methacrylmonomers based on 11-methacryloyl-amino-undecanoic acid which could be esterified with different alcohol's and ω-hydroxy-fatty-acids. A substrate of interest is enzymatically degradable cyclodextrin as ring component in polyrotaxanes. Cholic acid is a functional monomer that was incorporated into side chains of comb-like polymers via lipase catalyzed esterification in a regiocontrolled manner. Aspartic acid as a trifunctional amino-acid was shown to be suitable for the construction of polymerizable dendrimers. Aminochalcones were shown to be suitable for oxidative polycondensation by use of *horseradish* peroxidase and hydrogen peroxide.

Introduction
According to our interest in the chemistry of various types of functionalized polymers we intended to evaluate the potency of isolated enzymes as catalysts for the chemo-enzymatic construction and modification of methacryl monomers and synthetic polymers.

Several enzymes are meanwhile extensively applied as selective catalysts in low molecular weight organic chemistry [1]. The realization that enzymes can act not only in aqueous media, but also in dry organic solvents has opened a major field of enzymatic catalysis, e.g. monoacylation of sugar derivatives [2-3] or esterification by enzymatic condensation [4-7]. This means that, for example, also acrylmonomers [8-11] and even high molecular weight compounds which are only soluble in organic solvents can become potential substrates for enzymes [12-16].

Results and discussion
The interest of the present work is directed on the chemoenzymati modification of synthetic monomers and functionalized polymers by use of suitable enzymes as catalysts. Functional groups can be incorporated into polymers not only within the polymer main- chain, as usual, but they can also be combined in different manner with a branched or comb-like polymer (Scheme 1: A-D). Typically, comb-like polymers contain a linear main chain of different chemical structures and at least 4

C-atoms in the side chains. Comb-like polymers were investigated in the past general because of their interesting thermal and solution properties. However, up to now, only a relative moderate knowledge exists about the chemical and biochemical characteristics of functionalized comb-like polymers.

In this connection, we recently synthesized a new type of comb-like polymers bearing non covalently anchored cyclodextrins threaded onto linear and branched side chains of different polymers bearing bulky endgroups (Scheme 1: E,F). It is interesting to note that the cyclodextrins are biochemically prepared in a large scale via enzymatically catalyzed degradation of helically structured amylose. This new class of polyrotaxanes is characterized by a supramolecular architecture. It was shown in several experiments that the cyclodextrins can be removed from the side chains of the polyrotaxanes by selective enzymatic degradation.

Cholic acid is an interesting functionalized substrate that is produced from cholesterol via oxidation reactions. This commercially available substrate was used to be incorporated chemoenzymatically into comb-like polymers.

Chalcone-Derivatives are plant ingredients. In the present study, aminoderivatives from simple chalcones were found to be interesting substrates for enzymatic polycondensations.

Finally, polymerizable chiral dendrimers have been recently constructed via stepwise condensation reactions of aspartic acid, which is an enzymatically prepared trifunctional building block (Scheme 1: G).

As indicated above, we recently investigated the chemoenzymatic synthesis of functionalized polymers with comb-like structures by lipase catalyzed esterification of 11-methacryloyl-amino-undecanoic acid with a series of alcohol's (isobutyl alcohol, cyclohexanol, menthol, cholesterol, testosterone) and subsequent radical polymerization of the obtained esters [18-19]. The synthesis of a radically polymerizable oligoester (figure 1) was performed successfully also in the presence of lipases. This oligocondensation was recently improved by using a lipase from *Candida antarctica*.

In this connection, we also described a regioselective esterification of a gluco-pyranoside- and glucose derivative with 11-methacryloyl-aminoundecanoic acid in the presence of a lipase from *Candida antarctica* [20].

Cholic acid ($3\alpha,7\alpha,12\alpha$-trihydroxy-5β-cholan-24-oic acid) is an biologically active substrate that is produced from cholesterol via multienzymatic modification mainly in the liver tissue (scheme 2).

As a starting material for a technical synthesis of cortisone derivatives, cholic acid is isolated from ox or sheep bile in a large scale. According to scheme 2, cholic acid (**1**) contains, apart from the carboxylic group, three OH-groups in the 3-, 7- and 12-positions of the steroid component. Obviously, only one of the OH-functions is in the equatorial 3-position and therefore relatively more reactive than the sterically higher protected OH-groups in the axial 7 and 12-positions.

To evaluate the reactivity of the OH groups, a solution of cholic acid in THF was treated with a lipase from *Candida antarctica* as catalyst for several days. Regarding to SEC- and NMR-measurements, the isolated product was shown to be an oligoester (**6**). Under similar conditions, the oligocondensation in the presence of

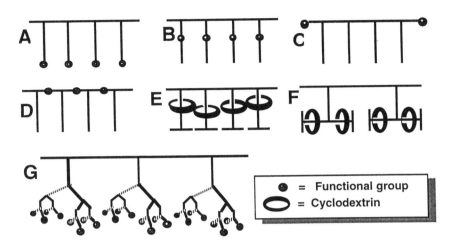

Scheme 1:
Illustration of polymer structures bearing functional groups

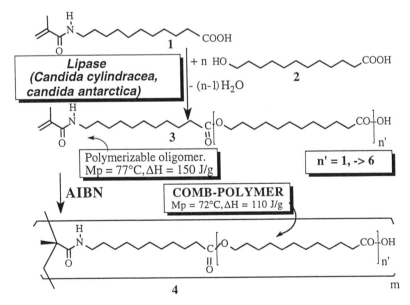

Fig. 1: Chemoenzymatic synthesis of a polymerizable oligoester

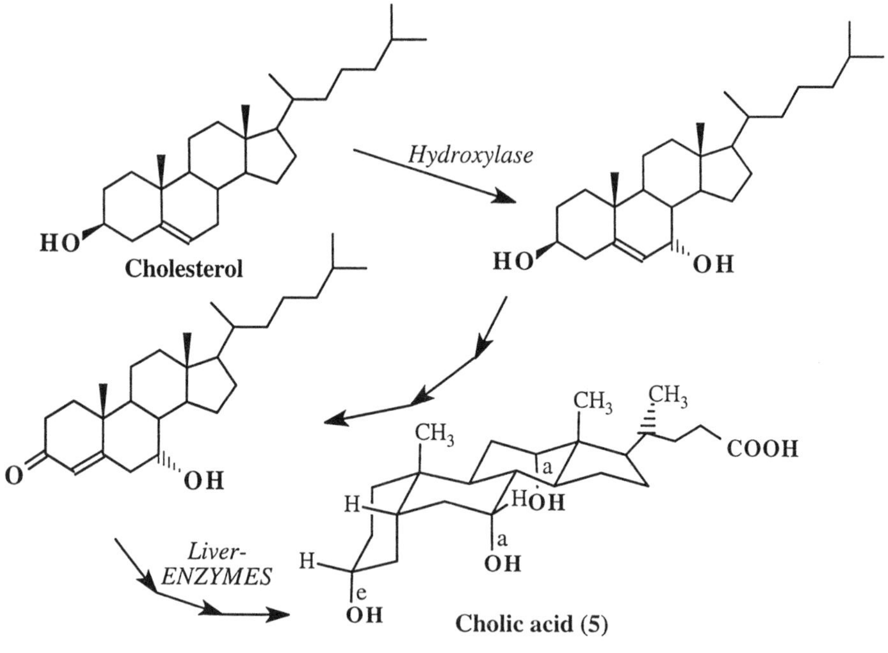

Scheme 2: Metabolism of cholesterol to cholic acid (5)

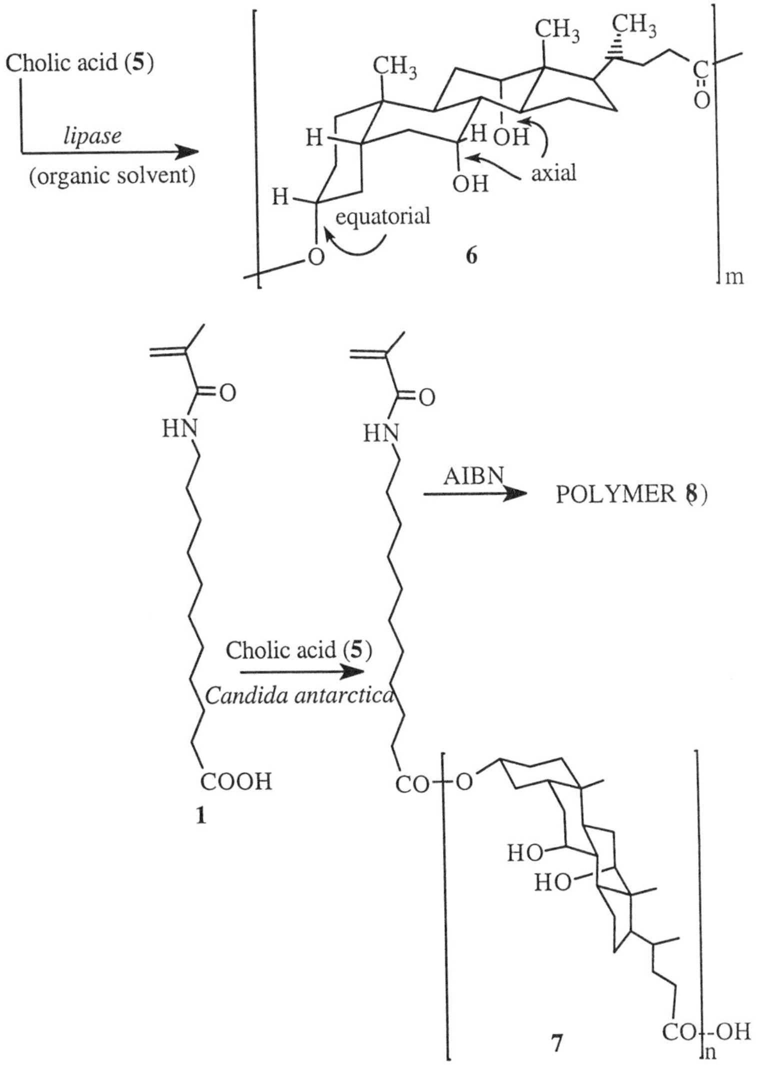

Fig. 2:
Esterification of cholic acid (5)

11-methacryloyl-amino-undecanoic acid (**3**) was performed yielding a polymerizable methacryl-monomer (**4**). Characteristic NMR-signals of cholic acid are compared with the corresponding signals of the esterified substrate. The spectra clearly show that the esterification of the carboxylic groups has occurred only with the OH group in equatorial 3-position. As mentioned above, this was expected from sterical viewpoint. The polymerization of the monomer **4** was realized in the presence of AIBN as initiator yielding the polymer (**5**). The structure of this methacrylpolymer was proved by NMR-spectroscopy showing characteristic signals as expected from the monomer (**4**). It can be concluded that cholic acid is an interesting tetrafunctional substrate that can be incorporated in synthetic polymers by use of lipase.

The use of cyclodextrins for the modification of monomers and polymers is a fascinating and rapidly developing area in polymer chemistry. Focusing the application of enzymes, the enzymatic ring-degradation of rotaxane-type substrates has been found in the case of the methacrylmonomer **8** (fig 3, A) and a comb-like polyketone (**9**) (figure) (fig 3, B). Up to now, the latter findings were qualitatively independent on the chemical structure of the polymer main-chain.

L-Aspartic acid can be produced in (fig. 4) technical scale from fumaric acid and ammonia by use of Lyase as enzymatic catalyst. In the present study, aspartic acid was stepwise condensed by use of classical peptide chemistry yielding a highly functionalized monomer **11**.

It was already published that unsubstituted aniline or m-phenylene diamine can be polycondensated by use of peroxidases as catalyst [17]. Recently, we succeeded to polycondensate aniline-derivatives containing photosensitive chalcone functions (**13**) by use of peroxidase (HRP) and hydrogen peroxide. From spectroscopic characterization, it was assumed that the structure of the polycondensate (**14**) is a result of amino-vinyl linkage as shown in fig. 5. It was shown that soluble oligomers are obtained at the beginning of the polycondensation while after a few days, a highly crosslinked material is yielded.

It was also shown that electron pour aniline-derivatives such as fluorinated chalcones, do not undergo this type of polycondensation reactions. Finally, a cyclodextrin complex of the chalcone (13) could not yet be polymerized under the same conditions.

Concluding remarks
The still ongoing investigations in the area of enzymatic catalysis in polymer chemistry have already shown the great potency of special enzymes for e.g. oligomerization and polymer modification reactions. However, many of the above described reactions could be performed, in principle, also by use of classical chemical methods. The main advantage of enzymatic catalyzation is that they are activity under mild conditions. Thus, in some cases they may help to prevent undesired side reactions. A further advantage of the enzymatic catalysis could be the stereocontrol of special chemical reactions. This means that it is still of great interest to evaluate the capacity of enzymes, even in the area of synthetic polymer chemistry.

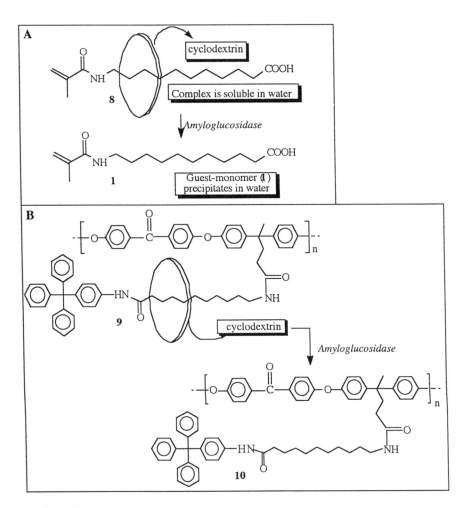

Fig. 3:
Degradation of cyclodextrin from a monomer (A) and from a polyrotaxane (B)

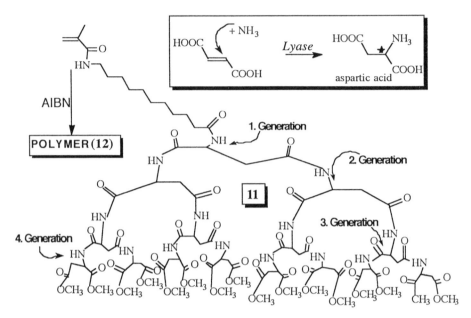

Fig. 4: Use of enzymatically prepared aspartic acid for synthesis of polymerizable dendrimers

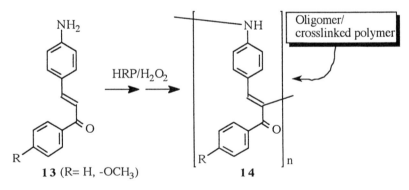

Fig. 5: Oxidative polycondensation of a amino chalcone in the presence of horseradish peroxidase (HRP)

References

01. C.H. Wong, G.M. Whitesides, Tetrahedron Organic Chemistry Series, Vol. 12, 1994, ElsevierScience, Ltd, Ed. J.E. Baldwin, FRS & P D Magnus, FRS, Pergamon
02. S. Riva, J. Chopineau, A.P.G. Kieboom, A.M. Klibanov; J. Am. Chem. Soc. 1988, 110, 584-589
03. M. Therisod, A.M. Klibanov; J.Am. Chem. Soc. (1986) 108, 5683-5640
04. I. Ikeda, A.M. Klibanov, Biotech. Bioeng., 42, 7881 (1993)
05. X. Chen, A. Jonathan, S. Dordick, D.G. Rethwisch, Macromol. Chem. Phys., 195, 3567 (1994)
06. D. R. Patil, J.S. Dordick, D.G.Rethwisch, Macromolecules, 24, 3462 (1991)
07. S. Matsumura, H. Kubokawa, K. Toshima, Makromol. Chem., Rapid Commun., 14, 55 (1993)
08. H. Ritter, C. Siebel, Makormol. Chem., Rapid Commun., 6, 521 (1985)
09. A. Ghorare, G.S. Kumar, J. Chem. Soc., Chem Commun. (1990), 134-135
10. A.L. Margolin, P.A. Fitzpatrick, P.L. Dublin; J.Am. Chem. Soc. 113, (1991) 4639-4694
11. I. Ikeda, J. Tunaka, K. Suzuki; Tetrahedron Lett. 32 (1991) 6865
12. H. Ritter; Trends in Polym Sci. (TRIP) 1 (6), (1993), 171
13. H. Uyama, K. Tanaka, S. Kobayashi; Proc. Japan Acad., 69. Ser. B (1993)
14. H. Uyama, S. Kobayashi; Chem. Lett. (1994), 1687
15. H. Kurioka, I. Komatsu, H. Uyama, S. Kobayashi; Makromol. Rapid Commun. 15 (1994), 507
16. O. Karthaus, S. Shoda, H. Takano, K. Obata, S. Kobayashi; J. Chem. Soc. Perkin Trans.1 (1994), 1851
17. J. H. Lee, R. M. Brown, S. Kuga, S. Shoda, S. Kobayashi; Proc. Natl. Acad. Sci. USA 91 (1994), 7425
18. K. Pavel, H. Ritter; Makromol. Chem. 192 (1991), 1941
19. K. Pavel, H. Ritter; Makromol. Chem. 194 (1993), 3369
20. U. Geyer, D. Klemm, K. Pavel, H. Ritter; Makromol. Chem., Rapid Commun., 16, 337-341 (1995)
21. M. Born, T. Koch, H. Ritter, Macromol. Chem. Phys. 196, 1761-1767(1995)
22. O. Noll, J. Jeromin, H. Ritter, paper submitted
23. G. Draheim, H. Ritter, Macromol. Chem. Phys., 196, 2211-2222 (1995)
24. M. Niggemann, H. Ritter, Acta Polymerica, 1996, in press,
25. S. Kobayashi, S. Shoda, H. Uyama, Adv. Polym. Sci. 121, 1 (1995)
26. Ch. Goretzki, H. Ritter, Macromol. Chem. Phys.(1996), in press

Author Index

Akkara, Joseph A., 112, 125, 167
Ashby, R. D., 175
Ayyagari, Madhu, 112
Banerjee, Sukanta, 125
Barbari, Timothy A., 188
Beckman, Eric J., 18
Beppu, Hideki, 74
Bisht, Kirpal S., 90
Brasseur, R., 144
Bruno, Ferdinando F., 167
Chaubal, Mahesh V., 188
Chaudhary, Apurva K., 18
David, C., 144
Day, Donal F., 175
Dordick, Jonathan, 2, 167
Gross, Richard A., 2, 90
Henderson, Lori A., 90
John, Vijay T., 125

Kaplan, David L., 2, 90, 112, 125, 167
Kobayashi, Shiro, 58
Lee, J. W., 175
Lefebvre, F. G. O., 144
Lenhart Joseph L., 188
Matsumura, Shuichi, 74
McPherson, Gary, 125
Payne, Gregory F., 188
Ramannair, Premchandran, 125
Ritter, Helmut, 199
Russell, Alan J., 18
Svirkin, Yuri Y., 90
Swift, Graham, 2, 90
Toshima, Kazunobu, 74
Uyama, Hiroshi, 58
Vanhaelen, M., 144
Wu, Katherine, 125

Affiliation Index

Faculté Universitaire de Gembloux, 125
Keio University, 74
Kyoto University, 58
Louisiana State University, 175
Rohm and Haas Company, 2, 90
The Johns Hopking University, 188
Tufts University, 2, 90, 112, 167
Tulane University, 125
U.S. Army Soldier Systems Command, 112, 125, 167

Université Libre de Bruxelles, 144
University of Iowa, 2, 167
University of Maryland Baltimore County, 188
University of Massachusetts—Lowell, 2, 90
University of Pittsburgh, 18
University of Wuppertal, 199

Subject Index

A–B-type condensation, enzymatic polyester synthesis, 20–21
AA–BB-type condensation, enzymatic polyester synthesis, 21–22
Acylation of alcohols, role of enzymes, 167–174
Alginates
  biocatalytic modification, 175–186
  conformations, 175–176
  enzymatic synthesis, 12
  sources, 175
Aliphatic polyesters, enzymatic synthesis, 64
Aminochalcone oxidative polycondensation, use of enzymes, 204, 206*f*
Amylose, regioselective enzymatic transesterification in organic solvents, 167–174
Aniline derivatives, enzymatic synthesis, 71, 72*f*
Applications, enzymes, 2–3
Aspartic acid stepwise condensation, use of enzymes, 204, 206*f*
Aureobasidium species, synthesis of poly(malic acid), 75

Benzyl β-malolactonate, enzymatic ring-opening polymerization, 74–87
Biocatalytic approach, novel phenolic polymers and composites, 125–143
Biocatalytic modification of alginates
  experimental description, 176
  experimental materials, 176
  experimental procedure
    acetylation of seaweed alginate by *Pseudomonas syringae* cells, 179, 182*f*
    culture conditions, 176–177
    deacetylation of bacterial alginate, 177
    molecular weight, 178

Biocatalytic modification of alginates— *Continued*
  experimental procedure—*Continued*
    NMR spectrometry, 177–178
    precipitation of alginates by metal ions, 178
    purification of acetylated alginates, 177
    quantitation of alginate and acetyl, 177
    viscosity, 178
  gelation with cations, 183, 185–186
  previous studies, 176
  separation of acetylation, 179
  viscosity effects, 183, 184*f*, 185–186
Biodegradable polymers, applications, 144
Biological catalyst use in polymer manufacturing, approaches, 188
Biomimetics, use of enzymes, 3
Bis(enol ester)s, enzymatic polycondensation with glycols, 64
Bis(2-ethylhexyl) sodium sulfosuccinate, structure, 127, 128*f*
tert-Butylcatechol, role in chitosan modification, 194
γ-Butyrolactone, structure, 91

Candida rugosa, use in synthesis of polyesters, 9–11
ε-Caprolactone, structure, 91
Cations, gelation, 183, 185–186
CdS-containing polymers, synthesis, 134–136
Cellulase, use in synthesis of polysaccharides and proteins, 11–12
Chemoenzymatic synthesis and modification of monomers and polymers
  aminochalcone oxidative polycondensation, 204, 206*f*
  aspartic acid stepwise condensation, 204, 206*f*

Chemoenzymatic synthesis and modification of monomers and polymers—*Continued*
cholic acid
esterification, 200, 203–204
production, 200, 202
cyclodextrin degradation, 204, 205*f*
experimental description, 199
functional group incorporation, 199–201
polymerizable oligoester synthesis, 200, 201*f*
Chitin, synthesis using enzymes, 12
Chitosan, modification by tyrosinase, 188–197
Cholic acid, use in modification of monomers and polymers, 200–204
Comb polymers, enzymatic synthesis, 14
Complex architectures, enzymatic synthesis, 14–15
Computer simulation, hydrolysis of dimethyl ester of *N*-succinyl-phenylalanine, 144–168
Cyclic acid anhydride, enzymatic poly(addition–condensation), 66, 67*f*
Cyclic acid anhydride polymer, enzymatic synthesis, 66, 67*f*
Cyclodextrin(s)
enzymatic degradation, 204, 205*f*
regioselective enzymatic transesterification in organic solvents, 167–174

Degradation of polymers, use of enzymes, 14, 36
Dendrimer formation, enzymatic synthesis, 14
Dihydroxyphenylalanine, enzymatic modification, 196–197
Dimethyl ester of *N*-succinyl-phenylalanine, hydrolysis, 144–163
Direct titration, analysis of enzymatic polyester synthesis, 31
Divinyl adipate, use in polymerization reactions, 25–27
Divinyl esters, enzymatic copolymerization, 64–66
Enantiomeric ratio, determination, 94

Enantioselective and regioselective transformations, role of enzyme suspensions, 91
Enol esters, use as acylating agents, 25
Enzymatic copolymerization
divinyl esters, 64–66
glycols, 64–66
lactones 64–66
Enzymatic modification of chitosan by tyrosinase
chitosan
film preparation, modification, and analysis, 190, 192–194
production, 188
structure, 189
potential modifications of chitosan, 194–197
previous studies, 188
reaction, 189–190, 191*f*
Enzymatic poly(addition–condensation), glycols and cyclic acid anhydride, 66, 67*f*
Enzymatic polycondensation, bis(enol ester)s with glycols, 64
Enzymatic polymerization of polyester and polyaromatic synthesis
applications, 71–72
polyaromatics
aniline derivatives, 71, 72*f*
polyphenols, 66, 68, 70–71
poly(phenylene oxide), 70, 71*f*
polyesters
aliphatic polyesters, 64
cyclic acid anhydride polymer, 66, 67*f*
lactones, 59–64
polyester copolymers, 64, 66
polyester macromonomer, 64, 65*f*
telechelics, 64, 65*f*
use of immobilized lipase, 66
Enzymatic polyphenol synthesis, feasibility, 125–127
Enzymatic reactions, importance, 58
Enzymatic ring-opening polymerization
four-membered lactones
experimental description, 75
experimental materials, 75–76
experimental procedure, 76–77
poly(β-malic acid), 83, 85–87

Enzymatic ring-opening polymerization—
   Continued
 four-membered lactones—*Continued*
   polymerization procedure, 76–77
   poly(β-propiolactone), 76, 78–84
   previous studies, 75
   lactones, 59–64
Enzymatic synthesis
 functional phenolic polymers with
   luminescent properties
   applications resulting from polymer
     morphology, 136–140
   experimental description, 127
   future directions, 138, 141, 142*f*
   polymer–semiconductor
     nanocomposites, 134–136
   polymer(s) with luminescent
     chromophores, 131–134
   safety considerations, 143
   surfactant C=O stretch, 127, 129*f*
 poly(phenylene oxide), 70
Enzymatic transesterification of
   polysaccharides in organic solvents, *See*
   Regioselective enzymatic
   transesterification of polysaccharides in
   organic solvents
Enzyme(s)
 polyester synthesis
   analytical methods, 28–31
   approach, 20
   design, 23–27
   end group analysis, 33, 37
   enzyme choice, 27
   enzyme stability recyclability, 50, 52, 53*f*
   future directions, 54
   kinetic features, Flory's analysis,
     52, 54
   kinetics model, 54
   microorganisms, 19–20
   molecular weight, evolution, 32–33,
     34–36*f*
   increases, 43–47
   routes, 20–22
   solvent, 31–32, 50, 51*f*
   stages, 37–43
   temperature effect, 50

Enzyme(s)—*Continued*
 polyester synthesis—*Continued*
   polymer science
   advantages, 125
   applications, 2–3
   features, 2
   key factors
     control of polymer structure, 4
     ease and flexibility of reactions, 4–5
     green chemistry, 5
   polymer degradation, 14
   polymer synthesis
     complex architectures, 14–15
     polyaromatics, 12–14
     polyesters, 9–11
     polysaccharides, 11–12
     proteins, 11–12
   sources
     alteration, 7–8
     enzyme engineering, 6
     isolation of novel enzymes, 6–7
   role in solvent–polymer interactions in
     molecular weight control of poly(*m*-
     cresol) synthesized in nonaqueous
     media, 112–124
   use in synthesis of polyphenols, 112–124
Enzyme engineering, source of enzymes, 6

Four-membered lactones, enzymatic ring-
   opening polymerization, 74–87

Gel permeation chromatography, analysis
   of enzymatic polyester synthesis, 28
Glycols
   enzymatic copolymerization, 64–66
   enzymatic poly(addition–condensation),
     66, 67*f*
   enzymatic polycondensation with
     bis(enol ester)s, 64
Green chemistry, use of enzymes, 5

^{1}H-NMR spectrometry, analysis of
   polyester synthesis using enzymes, 28
High-molecular-weight polycarboxylates,
   properties, 74–75
Horseradish peroxidase
   activity, 114

Horseradish peroxidase—*Continued*
oxidative polymerization of phenol
   derivatives, 66–70
use in synthesis and modification of
   monomers and polymers, 199–206
Hydrolysis of dimethyl ester of *N*-
succinylphenylalanine
computer simulation
   conformation of enzyme–substrate
      Michaelis complex, 154, 156–159
   design rules, 151
   geometrical optimization, 154, 155*t*
   initial papain structure, 151
   initial substrate structure, 151,
      152–153*f*
   potential energy of complexes, 159–163
experimental procedure
   computer simulation, 148–150
   hydrolysis, 148
   synthesis, 148
   kinetics, 150–151, 152*t*
   mechanism, 145–147
   substrate structure, 145
(Hydroxylethyl)cellulose, regioselective
   enzymatic transesterification in organic
   solvents, 167–174
(*p*-Hydroxyphenoxy)acetic acid, role in
   chitosan modification, 194–196
(4-Hydroxythio)phenol-containing
   polymers, synthesis, 134–136

Immobilized lipase, preparation, 66

Kinetics, hydrolysis of dimethyl ester
   of *N*-succinylphenylalanine, 144–168

Lactones
   enzymatic copolymerization, 64–66
   enzymatic polymerization, 59–64
   enzymatic ring-opening polymerization,
      59–64
   four-membered enzymatic ring-opening
      polymerization, 74–87
Lipase catalysis
   ring-opening polymerization
      four-membered lactones, 74–87
      lactones, 59–64

Lipase catalysis—*Continued*
   ring-opening polymerization—*Continued*
      ring-opening reactions for monomer
         and polymer synthesis
      advantages, 92–93
      chemoenzymatic routes to polyesters,
         93–97
      enzymatic stereoselective ring-opening
         polymerization, 97–101
      experimental description, 92–93
      lipase-catalyzed ε-caprolactone ring-
         opening polymerization, 102–108
      previous studies, 91
      synthesis of polyesters, 9–11
Luminescent property containing
   polymers, biocatalytic approach to
   synthesis, 125–143

Macrolides, enzymatic polymerization,
   59–64
Matrix-assisted laser desorption–
   ionization–MS
   description, 28, 31
   limitations, 31
α-Methyl-β-propiolactone, structure, 91
β-Methyl-β-propiolactone, structure, 91
Modification
   alginates, *See* Biocatalytic modification
      of alginates
   polymers, studies, 167–206
Molecular weight control of poly(*m*-
   cresol) synthesized in nonaqueous
   media, solvent–enzyme–polymer
   interactions, 112–124
Monomers
   chemoenzymatic synthesis and
      modification, 199–206
   synthesis using lipase-catalyzed ring-
      opening polymerization, 91–108
MS, analysis of enzymatic polyester
   synthesis, 28, 31
Mushroom tyrosinase, modification
   of chitosan, 188–197

Nonaqueous media, solvent–enzyme–
   polymer interactions in molecular
   weight control of poly(*m*-cresol),
   112–124

One-shot synthesis, polyester
macromonomer and telechelics
64, 65*f*
Optically active polymers, synthesis using
porcine pancreatic lipase, 24–25

Papain, role in hydrolysis of dimethyl
ester of *N*-succinylphenylalanine,
144–168
Penicillium cyclopium, synthesis of
poly(malic acid), 75
Peroxidase, use in synthesis of
polyaromatics, 12–14
Peroxidase-catalyzed oxidative
polymerization, phenol derivatives,
66–70
Phenol derivatives, peroxidase-catalyzed
oxidative polymerization, 66–70
Phenolic polymers and composites,
biocatalytic approach to synthesis,
125–143
Photosensitive chalcones, enzymatic
synthesis, 14
Physarum polycephalum, synthesis of
poly(malic acid), 75
Polyamides in the presence of papain,
hydrolysis, 144–168
Polyaromatics, enzymatic synthesis,
12–14, 58–72
Poly(*m*-cresol) synthesized in nonaqueous
media, solvent–enzyme–polymer
interactions in molecular weight
control, 112–124
Polyester(s)
biological synthesis
analytical methods, 27–31
approach, 20
design, 23–27
end group analysis, 33, 37
enzyme choice, 27
enzyme stability/recyclability, 50,
52, 53*f*
future directions, 54
kinetic features, 52, 54
microorganisms, 19–20
molecular weight, 32–33, 34–36*f*, 43–47
routes, 20–22

Polyester(s)—*Continued*
biological synthesis—*Continued*
solvent, 31–32, 50, 51*f*
stages, 37–43
temperature effect, 50
chemoenzymatic routes, 93–97
description, 18, 26*t*
reasons for interest, 74
synthesis
methods, 18
use of enzymes, 9–11, 18–54, 58–72
traditional synthesis
catalyst, 19
conditions, 19
Polyester copolymers, enzymatic
synthesis, 64, 66
Polyester macromonomer
enzymatic synthesis, 64, 65*f*
one-shot synthesis, 64, 65*f*
Poly(4-ethylphenol)
scanning electron micrograph,
127, 130*f*
synthesis, 141, 142*f*
Poly(glycolic acid), enzymatic synthesis,
10–11
Poly(β-hydroxyalkanoates)
description, 92
synthesis, 19–20
Poly(3-hydroxybutyrate), enzymatic
synthesis, 20
Poly(3-hydroxybutyrate-*co*-3-
hydroxyvalerate), enzymatic synthesis,
20
Poly(malic acid)
biodegradability, 75
lipase-catalyzed polymerization
debenzylation, 86
microbial degradability,
86, 87*f*
process, 83, 85
time course, 86, 87*f*
synthesis
methods, 75
use of enzymes, 10–11
Polymer(s)
chemoenzymatic synthesis and
modification, 199–206

Polymer(s)—*Continued*
role in solvent–enzyme interactions in molecular weight control of poly(*m*-cresol) synthesized in nonaqueous media, 112–124
synthesis using lipase-catalyzed ring-opening polymerization, 91–108
with luminescent chromophores, enzymatic synthesis, 131–134
Polymer modification, studies, 167–206
Polymer science, enzymes, 2–14
Polymer–semiconductor nanocomposites, enzymatic synthesis, 134–136
Polymer structure, control using enzymes, 4
Polymerization
enzymatic, *See* Enzymatic polymerization
stages, 37–43
(*R*)-Poly(α-methyl-β-propiolactone), structure, 92
(*R*)-Poly(β-methyl-β-propiolactone), structure, 92
Polyphenol(s)
applications, 112
enzymatic production
advantages, 112
methods, 112
structure, 112, 113*f*
synthesis, 12–13, 66, 68
production methods, 112
Poly(phenol oxidase), use in synthesis of polyaromatics, 14
Polyphenol particles, enzymatic synthesis, 70–71
Poly(phenylene oxide), enzymatic synthesis, 70, 71*f*
Poly(β-propiolactone)
enzymatic synthesis, 10–11
lipase-catalyzed polymerization
enzyme concentration vs. polymer molecular weight, 83, 84*f*
process, 74
time course, 80, 82*f*
synthesis methods, 74
Polyrotaxanes, enzymatic synthesis, 14

Polysaccharides
enzymatic synthesis, 11–12
regioselective enzymatic transesterification in organic solvents, 167–174
Porcine pancreatic lipase, use in synthesis of polyesters, 9–11
β-Propiolactone, enzymatic ring-opening polymerization, 74–87
Proteins, enzymatic synthesis, 11–12
Pseudomonas fluorescens, use in synthesis of polyesters, 9–11
Pseudomonas syringae
modification of alginates, 175–186
synthesis of polysaccharides and proteins, 12

Reaction(s), use of enzymes, 4–5
Reaction environment alteration, source of enzymes, 7–8
Regioselective enzymatic transesterification of polysaccharides in organic solvents
degree of substitution, 170–172
enzymatic specificity, 174
experimental description, 167–168
experimental materials, 168
experimental procedure, 168
Fourier-transform IR spectra, 168–170
previous studies, 167
proposed structure, 173*f*, 174
Rhizomucor miehei, use in synthesis of polyesters, 9–11, 25
Ring-opening polymerization
See Enzymatic ring-opening polymerization of four-membered lactones
See Lipase-catalysis, ring-opening reactions for monomer and polymer synthesis

Semiconductor–polymer nanocomposites, enzymatic synthesis, 125–143
Solvent, role in enzymatic polyester synthesis, 31–32, 50, 51*f*

Solvent composition, role in enzyme–
polymer interactions in molecular
weight control of poly(*m*-cresol)
synthesized in nonaqueous media,
112–124
Solvent–enzyme–polymer interactions in
molecular weight control of poly(*m*-
cresol) synthesized in nonaqueous
media
ethanol content effect, 115, 117–118
experimental description, 114
experimental materials, 114
experimental procedure, 114–115
future work, 124
hydroxyl group substitution, 121, 123*f*
Lineweaver–Burke plots, 118, 120–121
polymerization reaction scheme, 113*f*, 115
previous studies, 114
process, 121, 122*f*
solvent composition
  vs. enzyme activity, 118, 119*f*
  vs. polymer solubility, 118, 119*f*
thermal characteristics of polymers,
115, 116*f*
Stoichiometry, enzymatic polyester
synthesis, 23–24
Substrate(s), role in enzymatic polyester
synthesis, 23

Substrate alteration, source of enzymes, 8
Subtilisin Carlsberg, role in regioselective
transesterification of polysaccharides in
organic solvents, 167–174
N-Succinylphenylalanine dimethyl ester,
hydrolysis, 144–163
Synthesis, studies, 9–163

Telechelics
enzymatic polymerization, 64, 65*f*
one-shot synthesis, 64, 65*f*
Temperature, role in enzymatic polyester
synthesis, 50
Transesterification of polysaccharides in
organic solvents, *See* Regioselective
enzymatic transesterification of
polysaccharides in organic solvents
Tyrosinase
modification of chitosan, 188–197
synthesis of polysaccharides and
proteins, 12

Viscosity, role in biocatalytic
modification of alginates, 183, 184*f*,
185–186

Xylans, enzymatic synthesis, 11–12